Fluidization XVII

Edinburgh, United Kingdom
21-25 May 2023

ISBN: 978-1-7138-8677-8

Printed from e-media with permission by:

Curran Associates, Inc.
57 Morehouse Lane
Red Hook, NY 12571

Some format issues inherent in the e-media version may also appear in this print version.

Printed with permission by Curran Associates, Inc. (2024)

For permission requests, please contact AIChE
at the address below.

AIChE
120 Wall Street, FL 23
New York, NY 10005-4020

Phone: (800) 242-4363
Fax: (203) 775-5177

www.aiche.org

Additional copies of this publication are available from:

Curran Associates, Inc.
57 Morehouse Lane
Red Hook, NY 12571 USA
Phone: 845-758-0400
Fax: 845-758-2633
Email: curran@proceedings.com
Web: www.proceedings.com

TABLE OF CONTENTS

Digital Future of Chemical Industry- Challenges and Opportunities

Authors

Cordiner, J. - Presenter, University of Sheffeld

Fluidisation processes in industry present significant opportunities and challenges. This type of reactor offers increased yield and the ability to produce at scale in continuous processes with longer residence times and which need increased surface area of contact between the particles and the fluidisation gas. The speaker has experience of operating large fluidisation processes and will present some of the challenges that perhaps are not obvious from a lab view of fluidisation. We will look at how modelling can help understand and improve our processes, the limitations of the current available models and some opportunities for further research. When we look at the challenges facing our world we have a responsibility if we can make a step change, how can we make processes more efficient and take less energy or energy from renewable sources. We will discuss the challenges and opportunities that fluidisation can provide and challenge researchers to look for step change improvements.

KEYNOTE: 100 Years of Scaling Up Fluidized Bed and Circulating Fluidized Bed Reactors

Authors

Chew, J. W. - Presenter, Nanyang Technological University, Singapore

Cocco, R., Particulate Solid Research, Inc.

The year 2022 marked the 100th anniversary of the commercialization of the fluidized bed reactor. In those years, many new processes have been developed, with many considered breakthrough technologies, replacing technologies that were no longer considered competitive. Fluidized beds have the advantages of superior heat transfer and the ability to continuously move solids during operation, along with a list of other attributes. As a result, processes spanning coal and biomass gasification, pyrolysis, fluidized catalyst cracking, acrylonitrile, polyethylene, oxychlorination, and polycrystalline silicon have prospered with the application of fluidized bed technology. Expect that list to continue with current efforts in chemical looping, plastic pyrolysis, methane pyrolysis, and propane dehydrogenation, to name a few.

During that same period, we saw engineering go from an incremental approach to a more fundamental one using sophisticated models and relevant cold-flow experimentation. That engineering work process is still changing. The onset of better computational platforms, models, and experimental techniques, combined with better statistical tools (e.g., machine learning), leads to better understanding of the underlying physics and enhanced control systems (e.g., artificial intelligence). This will allow for the application of fluidized bed technology using less capital and less operating costs, and also enable commercialization in less time.

Experimental Investigation on Motion Characteristics and Thermal Behaviour of Lumps in Gas-Solid Fluidized-Bed Reactors

Authors

Errigo, M. - Presenter, University College London

Materazzi, M., University College London

Lettieri, P., University College London - Torrington Place

Gas-solid fluidized-bed reactors are characterized by excellent mixing. This led to their widespread use in waste pyrolysis and gasification processes. Although undesirable, the formation of lumps in such applications cannot always be prevented. If these lumps form, they may hinder the temperature homogeneity that usually characterizes fluidized-bed reactors. This work aims to investigate the motion characteristics and thermal behaviour of such lumps as these can help determine how lumps impact the reactor operation. A particular focus is placed on the segregation behaviour, the dispersion coefficients and the heat transfer coefficient between lumps and fluidized beds. The fluidization velocity, the particle size of the bed material and the density ratio between the lump and the bed material are varied to study their effects on the lumpâ€™s motion characteristics and thermal behaviour. A down-scaled pseudo-2D fluidized bed operated at ambient temperature is used. X-ray imaging is used to track the lumpâ€™s position, and infrared imaging allows to determine its temperature evolution over time. Dispersion coefficients were found not to be affected by the lump density, while they increased with the bed material particle size and the fluidization velocity. The segregation behaviour was mostly influenced by the fluidization velocity. An optimal value of gas superficial velocity guaranteeing proper mixing, regardless of the lump density, was found. The heat transfer coefficient was found to be strongly related both to the fluidization velocity and to the particle size of the bed material, with an increase in the particle size causing a deterioration of the heat transfer. Values of fluidization velocities guaranteeing maximum heat transfer coefficients were also identified.

Detection of Defluidization in Fluidized Beds

Authors

Shimizu, T. - Presenter, Niigata University

Fueki, S., Niigata University

Watanabe, M., Niigata University

Li, L., Niigata University

Kato, H., Niigata University

A problem of biomass/waste combustion is formation of agglomerates followed by sedimentation of coarse particles, which finally leads to defluidization. The mechanism of the formation of agglomerates is reaction between bed material (silica sand) and alkali (K and Na) which forms glassy material with low melting point. One approach to avoid this problem is to replace a part of bed material to maintain the concentration of alkali low. To optimize the system to feed/withdraw bed material, monitoring of fluidization is necessary. This work proposes a new method to detect sedimentation of agglomerates at the bottom of fluidized bed. Pressure fluctuation signal and gas velocity are measured in nozzle-type or bubble cap-type gas distributor. The fluidization status at the exit of the nozzle/bubble cap affects pressure fluctuation and/or gas velocity. In this work, cold model experiments were carried out to simulate the defluidization caused by agglomeration. Steel balls were fed to the fluidized bed as simulated agglomerates. The change in pressure fluctuation signal as well as gas velocity in the nozzles with the accumulation were measured. The fluctuation of the pressure was analyzed in terms of frequency spectrum calculated by use of FFT. Difference in frequency spectrum was observed between defluidization zone and normal fluidization zone.

Experimental and Modelling Study of the Volatiles Distributor for Improving the Horizontal Gas Distribution in Fluidized Beds

Authors

Li, X. - Presenter, Statoil

Lyngfelt, A., Chalmers University of Technology

Linderholm, C., Chalmers University of Technology

Mattisson, T., Chalmers University of Technology

Chemical looping combustion of biomass is an important and promising CO_2 capture technology due to its inherent CO_2 separation. However, biomass has a high content of volatiles, and if the contact between volatiles and the bed material of oxygen-carrier is insufficient in the fluidized bed used for the fuel reactor, the unconverted combustibles require the costly oxygen polishing step. Therefore, a concept named volatiles distributor (VD) is proposed for achieving a more uniform volatiles distribution over the cross-section, providing better gas-solids contacting and improving the gas conversion in the fuel reactor. The concept could be useful for other solid fuel conversion processes in fluidized-beds.

An experimental study of the VD with different configurations in a circulating fluidized-bed (CFB) cold-flow model has been investigated under different fluidization conditions, which shows that the VD gives a more uniform lateral distribution under higher fluidization velocity. Also, higher volatiles flow gives a more even distribution along the VD. The single bubble regime, i.e. large single bubbles formed at the bottom, and multiple bubble regime, i.e. multiple bubbles with different sizes formed at the bottom, were investigated with the VD. It was found that the VD gives a more uniform lateral distribution in single bubble regime compared to the multiple bubble regime. The installation of internal baffles at the bottom of the VD was found to reduce the bottom air flowing into the VD, thus improving the uniformity of the horizontal distribution of the volatiles and further reducing the risk of volatiles slip below the lower edge of the VD.

A modelling study is being conducted based on the above experimental data with the purpose to better understand the fluid dynamics in the fluidized bed with the VD.

Agglomeration Potential of Tropical Residual Biomass - Investigation on the Interactions between Ash Elements and Quartz Sand

Authors

Guío-Pérez, D. C. - Presenter, Universidad Nacional de Colombia

Achury Ortiz, S.

Perilla Vargas, P. J., Universidad Nacional de Colombia

Rincon-Prat, S. L., Universidad Nacional de Colombia

Cramer, T. H., Universidad Nacional de Colombia

In agro-industrial countries, residual biomass may have a significant technical potential for use in productive chains and in the energy sector. The use of residual biomass in thermochemical conversion processes entails some challenges, mainly derived from the variability of physical properties and the presence of specific chemical elements. Fluidized bed combustors are a convenient alternative for the conversion of biomass due to their flexibility and high conversion efficiency. However, agglomeration can occur due to the interaction between the biomass ashes and the bed material. The literature identifies temperature and concentration of some chemical elements (Si, Ca, Cl, and K) as the parameters with highest impact on the agglomeration potential. Likewise, melt-induced, and coating-induced has been identified as the most important agglomeration mechanisms in biomass combustion.

This work presents an experimental study of the interaction between biomass ashes and quartz sand at different temperatures under fixed bed conditions, for three selected tropical biomasses: oil palm kernel, oil palm rachis and coffee husk. A first assessment of the agglomeration potential was done through agglomeration indexes available in literature and based on the ash composition found for each biomass. Thereafter, quartz sand and biomass ash mixtures in a mass relation 90:10 were heated up (at a rate of 10 K/min). The final temperature and time in the oven were varied: 700, 800 and 900 °C and 2, 4, 6 and 8 hours, respectively. The product obtained from each experiment was analyzed using stereomicroscopy, X-ray fluorescence (XRF), scanning electron microscopy and energy dispersive X-ray analysis (SEM-EDX). Agglomeration indexes and experimental results on SEM-EDX and XRF allow a better understanding of the agglomeration mechanisms.

Gas Lateral Mixing in a Fluidized Bed - the Effect of the Solid Particles

Authors

Guío-Pérez, D. C. - Presenter, Universidad Nacional de Colombia

Praisa, U., Eindhoven University of Technology

Pallarès, D., Chalmers University of Technology

Leckner, B., Chalmers University of Technology

The mixing of different components in the gas phase is to a great extent responsible for the extent to which homogeneous reactions occur in a fluidized bed reactor. With regards to the lateral directions, gas mixing is often described as a dispersive process governed by a coefficient, which has been object of previous studies. This work, which is an initial step in a more exhaustive study of the gas mixing in fluidized beds, presents an investigation of the impact of the presence of a solids phase on the lateral mixing of the gas phase. The present work uses gas (CO_2) tracing measurements in a fluidized bed (3.1 m in height and 0.45 m^2 in cross section) operated at ambient conditions, using 106 μm glass beads as bed material and a range of fluidization velocities yielding both bubbling and circulating flow conditions. The concentration of the tracer gas was sensed at up to 35 positions over the cross section of the unit at different vertical distances from the tracer injection. The tracer gas concentration maps obtained experimentally were used to fit a convection-dispersion model and to calculate in that way the gas lateral dispersion coefficient. Measurements performed under solids-free conditions are compared to those in presence of solid particles, where the solids inventory and the fluidization velocity were varied, yielding different levels of solids concentration and fluxes.

KEYNOTE: Faraday Waves in Granular Particles Subject to Combined Gas Flow and Vibration

Authors

Boyce, C. - Presenter, Columbia University

Guo, Q., Columbia University

Da, W., Columbia University

Wei, J., Columbia University

Spitler, C., Columbia University in the City of New York

Experiments show that an analog of the Faraday wave instability forms in granular particles subject to combined vertical vibration and vertical gas flow. As compared to vibration alone, combined gas flow and vibration induces waves at lower vibration strengths and allows waves to persist at higher granular layer heights. Results across different vibration and gas flow conditions for the onset of waves are collapsed by introducing a non-dimensional velocity which incorporates the effects of gas flow and vibration. Continuum numerical simulations using a recently developed granular rheology model are able to reproduce wave patterns in grains subject to vibration with and without gas flow. Simulation results show that the added drag force in cases with gas flow causes more uniform forces and coordination motion among particles throughout the layer, which allow waves to persist at higher layer heights when gas flow is used.

Stirrer Design for Improving Fluidization of Cohesive Powders

Authors

Wu, K. - Presenter, Delft University of Technology

Kamphorst, R., Delft University of Technology

Bakker, A., Delft University of Technology

Wagner, E. C., Delft University of Technology

M.H. Meesters, G., TU Delft

Van Ommen, J. R., Delft University of Technology

Fluidization is widely used for a variety of solid processing applications, in which interphase mixing and heat and mass transfer are paramount. Steady fluidization is easily achieved using sand-like powders, whereas it is rather challenging to aerate fine particles because of intrinsic cohesion. Substantial attractive forces between individual grains dominate the collective behaviour, leading to the formation of agglomerates and gas channels which undermine greatly gas-solids interaction. Imposing additional forces and energies to the system is, therefore, necessary to aid fluidization. Several approaches have been studied to boost the fluidization of fine powders [1]. In particular, continuous stirring is shown able to effectively suppress channel formation, as well as mitigate agglomeration [2]. In addition, stirring is in favour of scale-up practices in terms of capital cost, operation and effective area size. Nevertheless, stirrers of different geometries exhibit dissimilar impacts, and there are only limited attempts of the stirred fluidization of cohesive powder. The effects of different stirrer designs remain largely unknown.

In this work, we investigate the impacts of stirring on the fluidization of cohesive powders under varying stirrer designs and operating conditions. Their performance were evaluated based on a comprehensive comparison across stirrer blade length, number, torsion angle, thickness and geometry. X-ray imaging technique was used to monitor the system hydrodynamics, such as gas bubbling and channelling behaviour. We observed that stirring effectively enhances the fluidization of cohesive powder, and mitigates gas channel formation. The experiments show that volume-based stirrers outperform plane-based designs in assisting fluidization. The analysis conducted forms a useful reference to optimise the stirrer design for existing and future applications.

[1] J.R. van Ommen, J.M. Valverde, R. Pfeffer, Fluidization of nanopowders: a review, J. Nanopart. Res. 14 (2012) 737.

[2] S. Alavi, B. Caussat, Experimental study on fluidization of micronic powders. Powder technol., 157(1-3) (2005) 114-120.

Bubble Characteristics in a Fluidized Bed Stripper for Three Types of Internals

Authors

Issangya, A. - Presenter, Particulate Solid Research, Inc

Karri, S. B. R., Particulate Solid Research, Inc. (PSRI)

Knowlton, T. M., Particulate Solid Research, Inc.

Cocco, R., Particulate Solid Research, Inc. (PSRI)

Chew, J. W., Nanyang Technological University, Singapore

A variety of standard and proprietary baffles are used in commercial fluidized bed strippers. The most common ones are disk and donut trays, horizontal sheds (also called 'chevrons'), structured packings and horizontal subway gratings. Gratings have not commonly been used but are now attracting more interest because they are easier to install, easier to service, have higher open areas and are less prone to flooding. The functions of any stripper internal are to suppress bubble growth, enhance bubbles distribution over the entire column cross section, slow down the bubbles and/or lengthen the time solids spend in the stripper to enhance mass transfer across the phases. This paper discusses local bubble properties measured in a fluidized bed stripper for three types of internals: gratings, disk and donuts and horizontal sheds. The test material was FCC catalyst.

The Real-Life Experience of two Ph.D. Candidates in an ITN: Benefits and Challenges

Authors

Winkelmann, M. - Presenter, The University of Edinburgh

Saghafian Larijani, R. - Presenter, University of Twente

Starting a Ph.D. is not only an academic but often also a cultural challenge, particularly for candidates who are moving to a new country and cultural environment. Working in an international academic environment can be exciting, but it also comes with several challenges. This presentation will provide insights from two Ph.D. candidates who are part of an EU-funded Innovative Training Network (ITN) called â€Training in Upscaling Particle Systems: Advancing Industry across Length-scalesâ€ (TUSAIL), https://tusail.eu/.

TUSAIL is a collaborative project that involves 15 Ph.D. candidates across six different universities and nine industrial partners in the EU, UK, and Switzerland. Being part of such an ITN is quite a different Ph.D. path and requests to not only acquire technical skills for the research project, but also to acquire interpersonal skills to make such an international project work.

The speakers will present insights into the differences between classic Ph.D. paths and Ph.D.s defined within an ITN. They will also provide insights into the challenges of working in different academic and industrial work environments and how to benefit from them. Additionally, the speakers will give insights and some of the challenges they experienced while navigating cultural differences that arise when working as an international team.

In conclusion, this presentation provides insights into the challenges and opportunities of working in an international Innovative Training Network. Attendees, both Ph.D. candidates and seniors, will gain insights and practical tips to help everyone navigate in an international academic environment.

KEYNOTE: CFD-DEM Modeling of Gas-Solid Reacting Flow (rCFD-DEM): From New Approaches to Industry Applications

Authors

Shen, Y. - Presenter, University of New South Wales

Computational fluid dynamics-discrete element method (CFD-DEM) has become a promising method to understand and optimize dense gas-solid reacting systems at the particle scale, yet the accuracy and efficiency still need to be improved for industry applications. In this work, a reactive CFD-DEM (rCFD-DEM) was developed, combining flow hydrodynamics, heat transfer, chemical reactions. A smooth method and a coarse-grained method were integrated to achieve high resolution and high efficiency, respectively. Two case studies are used for demonstrating that these new developments indeed advance the CFD-DEM modelling of fluid-solid reacting flows.

Gasification of Refuse Derived Fuels in Bubbling Bed Reactors: Bed-Freeboard Coupled Models, Results and Validation

Authors

Sebastiani, A. - Presenter

Iannello, S.

Materazzi, M., University College London

Gasification is the thermochemical conversion of solid feedstock into syngas, which can be accomplished through several reactor designs. Feedstocks containing high ash and volatile fractions, such as refuse-derived fuel (RDF) produced from solid wastes are commonly treated in fluidized bed reactors, which are particularly suitable due to their flexibility with respect to feedstock and operating conditions. Differently from conventional fuels, waste is generated from a number of different sources and influenced by a wide range of factors that create a complex mixture of materials, resulting into an extensive variability in RDF composition and, therefore, process operating conditions.

The complexity of simultaneous reaction phenomena occurring during RDF gasification and the corresponding lack of real plant data to validate models pose a significant challenge for accurately describing and predicting process performance. Thus, experimentally validated models that can correlate the large variations in process operating conditions and feedstock compositions to products quality (including an accurate prediction of tars and VOCs contents) are highly beneficial and of great industrial interest.

In this study, two models were developed to better describe the kinetics and hydrodynamics of different reactor sections of an RDF gasification process. The fluidised bed zone was modelled on Matlab to include feeding, fuel devolatilization and gas-particles interactions. The freeboard zone was modelled using Ansys Chemkin as a non-isothermal plug flow reactor to describe composition and temperature profiles of the syngas product, including over 300 components. The two models were then combined and validated with experimental data from laboratory and pilot scale reactors. The resulting system models allow for an accurate prediction of the effects of the operating conditions, such as pressure, equivalent ratio and use of novel gasifiying agents (e.g. pure oxygen, steam or CO_2 and their combination), on yield and quality of product syngas over a wide range of feedstock compositions.

Polydisperse, Segregating Fluidized Beds: From Detailed, Small-Scale Investigations to Fast, Large-Scale Simulations

Authors

Lichtenegger, T. - Presenter, Johannes Kepler University

Pirker, S., Johannes Kepler University

Atzori, M., Johannes Kepler University

Fluidized beds with polydisperse particle size distributions can exhibit segregation, which might even lead to undesired, partial defluidization. Corresponding numerical simulations like CFD-DEM are challenging because of the distinct time scales involved: Grain collisions take place within microseconds whereas meso- and macroscopic mass transport happens within seconds. Finally, segregation might evolve over the course of minutes. Therefore, CFD-DEM simulations covering such durations are only feasible for small test cases below the spatial scales of many experiments or of industrial reactors. Nevertheless, such small-scale investigations can provide relevant insights. We studied the influence of the gas inlet profile varying between homogeneous aeration, a central spout and inflow at the sides of the bed for fixed total mass flux. Switching between the different profiles allowed us to control segregation in a simple fashion that may be easily implemented in experimental setups without the need to adapt the overall gas flow.

Furthermore, we discuss how fast, long-term simulations of larger scales may be carried out. We employ our data-assisted time-extrapolation technique recurrence CFD (rCFD), which provides a numerically cheap approximation for recurrent, fast degrees of freedom (such as the particle velocity field in a fluidized bed). On top of the resulting long time series of flow fields, passive or weakly coupled transport processes can be investigated in an extremely efficient way. Based on preliminary results, we are optimistic that rCFD simulations can be used to picture the long-term evolution of polydisperse fluidized beds, e.g. to understand the impact of continuous granulation, progressing segregation and the impact of counter-strategies like switching between different inlet profiles. Ultimately, real-time capable rCFD simulations might be used to monitor the behavior of an industrial reactor and help to control it.

A Fluidised Bed Pyrolysis Process for Chemical Recycling of Mixed Plastic Waste

Authors

Windows-Yule, K. - Presenter, University of Birmingham

Seville, J., The University of Birmingham

Leeke, G. A., University of Birmingham

Thorpe, R. B., School of Engineering

Griffiths, A., Recycling Technologies

Besong, M., Recycling Technologies

Sanchez, A., Recycling Technologies

Bhattacharya, S., Recycling Technologies

The ecological damage from plastic pollution is more evident by the day, but global manufacture and use is increasing rapidly. Even if substitutes are found in some limited cases, the unique advantages of plastic will continue to make it the material of choice in many applications. Recycling is essential and the good news is that effective processes are available now to enable this to happen. Fluidised bed pyrolysis is a leading candidate for chemical recycling to provide hydrocarbon feedstocks for reuse, but presents a number of practical problems in its implementation. This paper summarises the design and operation of a twin-fluidised bed process in which pyrolysis occurs in a bubbling bed reactor of sand-like material which is heated in a second bed by combustion of light hydrocarbon gas, with solids exchange between the two. The mixed hydrocarbon product is separated in a series of columns and clean-up stages. The process development is described, including the measurement of the pyrolysis kinetics, the design of the two fluidised beds and ancillary equipment and the overall design of an industrial scale pilot plant at Swindon, UK, which has successfully processed the equivalent of 700 tonnes/year in this way. Experience at this scale has been used in the design of a 7000 tonnes/year design. The vision is for a network of hundreds of such plants across each country distributed so as to minimise transport issues. Challenges include maximising the efficiency and availability of the process, ensuring that contaminant molecules are effectively removed and devising an industrial ecology for a circular economy which makes this approach sustainable in both environmental and economic terms.

Effect of CaO Additives on PET Steam Gasification in Bubbling Fluidized Bed

Authors

Li, S. - Presenter

Inayat, M., Aalto University

Jarvinen, M., Aalto University

The growth of polyethylene terephthalate (PET) consumption for beverage bottles, packaging and textiles and their ineffective recycling procedures globally causes serious environmental problems. Steam gasification could be one possible way to recover materials from end-of-life PET, and CaO could catalyze the hydrogen production. The objective of this research was to study the effect of CaO on PET the gaseous products from steam gasification experimentally. CaO/PET mass ratios 0, 1, 2, 3 and 4 were fed in batch into a bubbling fluidized bed reactor at 750°C. N_2 and steam were employed as fluidization gas and steam to PET mass ratio was 2.5. After the tar and water vapor removal by impinger bottles with isopropanol, the product gas was analyzed by an online syngas analyzer. The results showed that hydrogen yield doubled (from 10.6 mol/kg PET to 21.9 mol/kg) when CaO/PET ratio increased from 0 to 4. CO_2 yield also increased with increasing CaO/PET ratio instead of decreasing because $CaCO_3$ decomposition is more active at 750 °C. However, at higher CaO/PET ratios, the yield seems to reach a constant level at 15 mol/kg. CO, CH_4 and other non-condensable hydrocarbon gases varied slightly with CaO/PET ratios where only CO rose 2 mol/kg PET comparing with no CaO addition. The batch feeding, 1g/min PET and corresponding ratio CaO, is reflected by the instantly measured periodical peaks of product compositions. Similar amplitudes in each product indicate a relatively stable gasification process. Moreover, in each period, H_2 and CO_2 have opposite trends as CH_4 and CO which could be due to the water-gas shift reaction. The results could be applied for hydrogen production from end-life PET steam gasification.

Visualization Experimental Study on Hydrodynamic Characteristics of Gas-Liquid-Solid Mini-Fluidized Beds Based on PIV

Authors

Liu, M. - Presenter, State Key Laboratory of Chemical Engineering (Tianjin University)

Ma, Y., Tianjin University

Li, C., Tianjin University

Pyroclastic density currents (PDCs) are the most lethal volcanic process on Earth. Forecasting their inundation area is essential to mitigate their risk, but existing models are limited by our poor understanding of their dynamics. Here, we explore the role of evolving grain-size distribution in controlling the runout of the most common and deadliest PDCs known as block-and-ash flows (BAFs). Through a combination of theory, analysis of deposits and experiments of natural mixtures, we show that rapid changes of the grain-size distribution transported in BAFs result in the reduction of pore volume (compaction) within the first kilometres of their runout. We then use a multiphase flow model to show how the compressibility of granular mixtures leads to fragmentation-induced fluidisation (FIF) and excess pore-fluid pressure in BAFs. This process dominates the first ~2 km of their runout, where the effective friction coefficient is progressively reduced. Beyond that distance, transport is modulated by diffusion of the excess pore pressure. Fragmentation induced fluidisation provides a physical basis to explain the decades-long use of low effective friction coefficients used in depth-averaged simulations required to match observed flow inundation.

KEYNOTE: Novel Methods for the Efficient Calibration and Rigorous Validation for Simulations of Fluidised and Spouted Beds

Authors

Windows-Yule, K. - Presenter, University of Birmingham

Werner, D., University of Birmingham

Nicusan, A. L., University of Birmingham

Che, H., Institute of Engineering Thermophysics, Chinese Academy of Sciences

Seville, J., The University of Birmingham

Numerical methods such as the discrete element method (DEM), computational fluid dynamics (CFD), coupled CFD-DEM, and multiphase particle-in-cell (MP-PIC) simulations provide potentially powerful tools for the investigation and optimisation of fluidised beds and other multiphase systems. However, without rigorous calibration and validation, their outputs may be inaccurate, or even entirely unphysical. The process of calibration can, however, prove highly time- and labour-intensive, and both calibration and validation are sorely lacking a clearly-defined "Best Practice".

In this talk, we introduce novel methods for both the calibration and validation of numerical simulations of fluidised- and spouted-bed systems, both of which can be easily generalised to other particulate, fluid or multiphase systems, and to arbitrary simulation techniques.

In part 1 of the talk we introduce Autonomous Characterisation and Calibration using Evolutionary Simulation (ACCES), an open-source, AI-based tool which uses evolutionary computation and metaprogramming to autonomously determine the correct calibration parameters for a given simulation based only on data readily available from widely-available powder characterisation tools.

in part 2, we introduce Positron Emission Particle Tracking (PEPT), a nuclear imaging technique which allows the fully-three-dimensional dynamics of diverse particulate and multiphase systems to be imaged with sub-millimetre spatial resolution and sub-millisecond temporal resolution, even deep within the interior of large, dense, optically opaque systems. We show how the detailed, three-dimensional fields extracted from PEPT data may be used to provide a uniquely rigorous validation of simulation methods such as CFD, DEM and MP-PIC.

We demonstrate the application of both methodologies using a series of case studies involving both laboratory and industrial systems, including spouted-bed coffee roasters and fluidised-bed-based pyrolysis reactors for the chemical recycling of plastic waste.

Characterization and Modeling Method for Transient Macroscopic Gas-Solid Flow in Circulating Fluidized Bed Riser

Authors

Ritvanen, J. - Presenter, Lappeenranta University of Technology

Nikku, M., LUT University

Fluidized bed processes are common in the process and energy industries because they have very good and efficient reaction conditions. These conditions are based on very strong mixing and gas-solid hydrodynamics. The hydrodynamics of gas-solid suspension in fluidized beds is very complex and includes several coupled phenomena. Knowing hydrodynamics is key in process design and development. Many types of simulation methods with different accuracy have been used in the design of fluidized beds and in examining the operation of processes. In all previous investigations and simulation attempts, the main challenge has been determining the interaction between the phases, i.e., the magnitude of the drag force under different conditions. This work presents a practical characterization approach to determine the gas-solid drag force. In this method, the interaction is determined experimentally for real solid material, which includes size, shape, and density distributions. In addition, a macroscopic numerical simulation model is presented that can be used to examine time-dependent gas-solid flows. This macroscopic 1D gas-solid flow model is based on the momentum equation of the solid phase, including an experimentally characterized interaction between phases. The developed model is able to predict the steady-state and time-dependent gas-solid flow profiles in the circulating fluidized bed riser. The work presents a description of the interaction term between phases based on experimental results, as well as the validation of the developed flow model in the operating range corresponding to the experimental results. In addition, the flow model simulation results for time-dependent variation in the circulating fluidized bed riser are given, and the results are compared with experimental results.

A Novel Multi-Resolution Hybrid CFD-DEM Method for High-Fidelity Modelling of Cross-Scale Particulate-Fluid Flow

Authors

Xie, Z. - Presenter

Shen, Y., University of New South Wales

Cross-scale polydisperse particulate-fluid flows are widely practised yet their multi-resolution simulations are not achieved due to the lack of reliable numerical methods. In this work, a novel so-called hybrid CFD-DEM model is developed for the high-fidelity multi-resolution simulations of cross-scale polydisperse particulate-fluid flow systems, for the first time featuring no limit on the grid size to particle diameter ratio (l_m/d_p). The hybrid CFD-DEM method collaboratively considers the fluid-coarse particles interaction (l_m/d_p <0.1), fluid-medium particles interaction (0.1< l_m/d_p < 3), and fluid-fine particles interaction (l_m/d_p >3) by means of the FD (fictitious domain)-based resolved method, semi-resolved method, and unresolved method, respectively. The model effectiveness and accuracy are validated by the three typical polydisperse particulate-fluid flow scenarios. The results confirm the ability of the hybrid CFD-DEM model of capturing both the detailed flow fields of fluid and particles and particle-fluid interactions in the particulate-fluid flow with a cross-scale size distribution. Moreover, the hybrid CFD-DEM model shows excellent capability in mesh convergence, wide applicability, and high accuracy. This work provides a promising tool for the modelling of many cross-scale particulate-fluid flow systems.

A Hard Toolbox to Improve Supervision of PhD Studies

Authors

Pallarès, D. - Presenter, Chalmers University of Technology

Supervision of PhD studies is widely recognized as a difficult task demanding a number of skills which are difficult to master and are often gained through painful experience. In order to identify paths for a more efficient learning of the role as PhD supervisor, dedicated workshops involving both faculty and PhD students have been held in addition to more than 40 exit interviews with PhD graduates. This work has resulted in a series of tools for the diagnostics of PhD supervision which are currently being introduced as part of the regular follow-up of PhD studies at Chalmers University of Technology. The tools are hard in the sense that they involve quantitative indicators which are initially self-evaluated by the PhD student and the supervisor, and then later compared for common reflection and, if needed, agreement on action. Further, the results contain also data on mutual expectations between the PhD students group and the supervisors group, and comparison to the guessed expectations on each.

The presenter was awarded the prize *PhD Supervisor of the Year* at Chalmers University of Technology (Sweden), is deputy Head of Dept. for PhD studies, and participates actively in the Chalmers strategic group for pedagogical development of PhD supervisors.

On the Accuracy of the Energy Minimization Multiscale (EMMS) Model for Circulating Fluidized Beds

Authors

Bhat, S. - Presenter

Pakseresht, P.

Yao, Y., Engineering and Process Science, Core R&D, The Dow Chemical Company

Fan, Y., The Dow Chemical Co

Theuerkauf, J., The Dow Chemical Company

Capecelatro, J., Dept of Mechanical Engineering

Coarse grid simulations of circulating fluidized bed (CFB) reactors require accurate predictions of the particle drag force. Conventional homogeneous drag laws are known to be insufficient for such configurations owing to the heterogeneous structure of the particles that emerges (e.g., clustering). The Energy Minimization Multi-Scale (EMMS) model has received much attention for capturing the effect of heterogeneity in simulations of CFB reactors. Originally derived for calculating the net particle force in CFB risers, EMMS has been integrated within coarse grid simulations of Euler-Euler (EE) and Euler-Lagrange (EL) approaches to modify the drag force in each computational cell.

The EMMS approach decomposes the flow field into a dense and dilute phase that interact with each other. Many assumptions are made in the derivation of this model including (i) heterogeneity is represented by a single spherical cluster; (ii) invariant gas and particle superficial velocities for each phase; (iii) constant void fraction for each phase; and (iv) invariant particle drag force within the phases, among others.

Leveraging highly resolved EL simulations, we present a critical assessment of the EMMS model. The fundamental assumptions are probed in detail and model accuracy is quantified. We consider two flow regimes (i) fully developed settling of inertial particles under gravity and (ii) fully developed particle-laden pipe flow, representing unbounded and wall-bounded regimes, respectively. The former mimics the individual computational cells in the coarse-grained EE or EL simulations to which the EMMS model is typically applied, while the latter resembles the original application of the EMMS model for fast fluidization regimes. A structure tracking algorithm is applied to the EL data to measure cluster characteristics. Results show significant variation in the two-phase flow parameters, unlike the single-value predictions of the EMMS model. This work can provide a framework for incorporating the observed variation and improving the EMMS model.

KEYNOTE: Combining fluidization and gas-phase coating: surface modification at the nanoscale

Authors

Van Ommen, J. R. - Presenter, Delft University of Technology

Nanostructured particles (e.g., core-shell nanoparticles or porous micro-particles containing nanoparticles) have high potential in applications such as catalysis and energy storage, but also in medicine and food. However, to come from product design incorporating nanostructuring to actually making such materials at relevant scales is not trivial. Gas-phase coating using techniques such as atomic layer deposition (ALD) can be used to provide the surface of a particle with either an ultrathin continuous coating or a decoration of nanoclusters. In this way, we can benefit from the wide range of chemistries developed over the years for ALD.

When carried out in a fluidized bed or a pneumatic transport reactor, gas-phase coating is an attractive way of producing nanostructured particles with excellent scale-up potential. Even nanopowders can be fluidized to be modified this way, although they are not present as individual particles but as very dilute agglomerates.

I will discuss how to fluidize (ultra-)fine powders in order to obtain a high coating quality. Also other requirements for this coating approach will be examined. I will demonstrate how it can be deployed to tune the wettability behaviour of particles, and how it can be used to give pharmaceutical powders controlled-release properties. Moreover, the application to catalysis will be discussed.

Fluidization and Mixing Behavior of Nano and Micron Sized Particles in a Vibrated Fluidized Bed

Authors

Hua, Z. C. - Presenter, Hamburg University of Technology

Pietsch-Braune, S., Hamburg University of Technology

Heinrich, S., Hamburg University of Technology

All-Solid-State-Batteries (ASSBs) are set to provide high energy densities along with a safer energy storage medium than conventional batteries, which could become crucial in the electrification of transport systems. One of the main challenges of ASSBs is the upscaling of the production process [1]. To that end, the use of a vibrating fluidized bed is proposed for the mixing, coating and aggregation of the battery materials. Thus, the fluidization and mixing behaviour cathode material consisting of lithium iron phosphate (LFP), carbon black (CB) and sodium chloride (SC) is investigated in a vibrated fluidized bed with pulsated gas flow and a high velocity microjet. SC was used as a model material for solid electrolytes, which often react with moisture. Strictly vertical vibration was implemented using a linear shaft. The pulsated gas flow was implemented via solenoid valves. The microjet enhances the fluidization and also plays a major role in the creation of hetero agglomerates or aggregates that consist of multiple species.

A parametric study including vibration parameters, pulsated gas flow parameters and pre-conditioning of the particles via drying or sieving was explored while the fluidization characteristics were investigated via measurement of pressure drop and bed expansion. Lastly, microjet parameters, such as orifice diameter or jet-pressure, were varied. The success of the hetero-agglomerate and aggregate production was determined by SEM and TEM images in conjunction with particle size measurements. Additionally, the electrical powder conductivity of the produced aggregates and agglomerates were measured via a four-point probe method.

It was shown that the vibration-, jet- and the pulsated gas flow parameters influence the fluidization characteristics. Further, the mixing quality improved with extended fluidization time. The successful production of composite material was confirmed via electron microscopic techniques.

Systematic Analysis of Mixing and Segregation Patterns in Polydispersed Fluidised Beds for Combined Catalytic Applications

Authors

Zhu, H. - Presenter

Materazzi, M., University College London

Lettieri, P., University College London - Torrington Place

Fluidized beds are among the most promising technologies in renewable energy and chemical production, especially in bioenergy and biorefining systems, due to their excellent mixing features, good operating flexibility, and enhanced heat and mass transfer. In industrial operations, sizes and densities of the solid particles in a fluidised bed are various, and have significant influence on the fluidization process. The mixing or segregating behaviour of different classes of particles is essential for developing relevant chemical or physical processes and reactor designs, especially those where different catalysts or solid adsorbents are coexisting in the same reactor. However, limited research has been focused on the characterization of fluidization profiles of mixed particles in fluidized bed reactors, and at real operating conditions.

In this work, we investigate the expansion and segregations behaviours of mixed Geldart group powders (in binary or ternary systems), to mimic operation of polydispersed beds where different bed materials/catalysts are employed at the same time.

A novel X-ray imaging facility will be used to analyse and describe the real-time mixing and segregation profiles in a lab-scale reactor operated at different fluidization regimes and temperatures. The non-invasive X-ray technique can provide frame-by-frame imaging with extremely high time resolution, offering visual representation for the bed characterization within the reactor at different operating conditions. This study also aims to investigate the feasibility of combined catalytic applications in fluidised bed systems, such as twin beds and circulated fluidised bed systems. Knowledge generated in this study can help to design, improve, and optimise advanced thermochemical conversion technologies, especially those involving in-situ gas adsorption, such as sorption enhanced catalytic processes with catalyst cyclic regeneration, to overcome thermodynamic limitations of conventional processes, increasing efficiency and resilience.

Finding the Golden Threads: A History of (some) Women in Chemical Engineering

Authors

Baker, N. - Presenter

Ocone, R., Heriot Watt University

The profession of chemical engineering can, rightly, hold its head up high when it comes to demonstrating that more women seem to wish to enter its ranks, globally, than some other branches of engineering. This paper will summarise some of the statistics supporting this statement and will demonstrate how this has come about. From the history of chemical engineering undertaken by medieval â€˜alchemistsâ€™, such as Frenchwoman Marie Meurdrac (c.1610-80CE), to the latest projects in nanoscale chemical products, a selection of womenâ€™s biographies will be used to show how women have worked in different eras across the world.

Many of the earliest women in this paper were never described by the 20th century definition of â€˜chemical engineerâ€™ during their lifetimes, hence we define our examples by what they did rather than any labels or qualifications. It is notoriously difficult in researching womenâ€™s history to be sure of who was the â€˜First woman to do Xâ€™ but some such claims are examined. Do we count American Marion Rice Hart as the first female chemical engineering graduate (MIT, 1914) because that was the description of her degree? What about Romanian Elisa Leonida-Zamfirescuâ€™s degree in engineering with chemistry (Royal Academy of Technology, Berlin, 1912)?

Although there are extant studies of the histories of women in chemistry and in engineering, the authors believe that their project, from which these sample histories derive, is the first attempt at an in-depth research into the international history of women in chemical engineering. Perhaps some of the fascinating women we have discovered might become inspiring portraits hanging on academic walls alongside the aged paintings of long-gone men.

If engineering is to be a diverse field which women wish to enter, at all levels, it helps to have pride in a great history as well as a promising future.

Migration Characteristics of Fine Cohesive Particles in a Vibrating Gas-Solid Fluidized Bed

Authors

Mawatari, Y. - Presenter, Kyushu Institute of Technology

Nakamura, R., Kyushu Institute of Technology

Sonoda, K., Kuyushu Institute of Technology

Suizu, C., Kyushu Institute of Technology

This study investigated the flow pattern transition for fine cohesive particles in a gas-solid fluidized bed under mechanical vibration. In general, as the particle diameter decreases, the interaction force, such as van der Waals force, acting between particles becomes larger; the bed consists of various sizes and shapes of agglomerates that are not favored for fluidizing gas to flow through evenly. It is well known that mechanical vibration is useful to improve the fluidity of fine cohesive particles due to the breakage of the agglomerates and the stable gas channels by the addition of the vibration. In this study, the migration behavior under vibrating conditions was focused on to examine the interrelation between the vibration-induced particle migration behaviors and bed fluidity. When the vibration intensity was larger than a certain magnitude, the agglomerates started to form convective particle migration (titling or heap types depending on the vibration intensity). Once the agglomerates migrated, the bed pressure drop decreased and its decreasing degree became larger as the migration velocity of agglomerates increased. As the vibration intensity increases, it is considered that the compression by the vibration to the particle bed becomes larger, and the fluidizing gas would be difficult to percolate through the consolidated part. In such situation, the gas flow concentrates to make a preferential flow to migrate the agglomerates to the bed surface. This preferential gas flow induced by the vibration is considered to cause a decrease in the bed pressure drop. In this study, from the interrelation between the particle migration behaviors and the bed pressure drop was experimentally investigated to clarify the fluidization mechanism and the flow pattern transition for fine cohesive particles in a vibrating gas-solid fluidized bed.

KEYNOTE: Modeling and Harnessing Mesoscale Structures in Fluidization and Multiphase Reactors

Authors

Guan, X. - Presenter, Institute of Process Engineering, Chinese Academy of Sciences

Mesoscale phenomena are ubiquitous in multiphase reaction systems where the material, reactor and system levels all feature the mesoscale structures, such as molecular assemblies, particle (bubble/droplet) clusters, or reactor networks, respectively. Mesoscale problems are essential to a more fundamental understanding of momentum, mass and heat transfer in the classical study of transport phenomena, and to the mixing, residence time distribution and rate-limiting analysis in the chemical reaction engineering, yet they are beyond the scope of those classical textbooks of chemical engineering. In this presentation, I will highlight a heuristic mesoscale modeling approach for multiphase reactor systems, starting from a conceptual Energy-Minimization Multiscale (EMMS) model and ending at the stability-constrained multifluid CFD model. While the stability condition determines the direction of system evolution, the stability-constrained CFD further describes the dynamics of structure evolution. Then this theory is applied to the simulation of gas-solid fluidization and gas-liquid bubble column reactors, with further extension to gas-liquid-solid three phase flow and stirred tanks. Several industrial applications on meso-scale modeling in liquid-solid polyethylene reactors and liquid-liquid emulsification systems will also be highlighted.

Modeling Sorption Enhanced Methanation in Dual Interconnected Fluidized Beds with Different Sorbents

Authors

Scala, F. - Presenter, Istituto di Ricerche sulla Combustione, Consiglio Nazionale delle Ricerche

Coppola, A., P.le Tecchio 80, 80125, Napoli, Italy

Massa, F., Università degli studi di Napoli Federico II

In the framework of the energy transition, the possibility of combining the use of CO_2 from carbon capture techniques with a power-to-gas technology, exploiting the renewable electric energy surplus to produce synthetic methane from hydrogen, is gaining attention. Sorption enhanced methanation (SEM) may improve the performance of the process shifting the equilibrium composition towards a higher CH_4 production, enabling, for example, to reduce the operating pressure for methanation. The concept involves the removal of the produced steam using a selective sorbent. A novel dual interconnected fluidized beds (DIFB) SEM configuration was recently suggested to overcome fixed bed limitations such as the transient operation mode and the difficult temperature control.

CO_2 DIFB-SEM chemical looping was simulated in AspenPlus, investigating two possible sorbents: CaO derived from limestone and a commercial zeolite 3A. Two â€œFluidBedâ€� blocks, including all significant reaction kinetics, simulated the methanator in a range of temperature from 250Â°C to 350Â°C and the regenerator at 450 Â°C, where the spent sorbent is continuously recirculated for dehydration. The analysis was carried out focusing on the outlet gas quality in terms of contaminants for the direct injection in the natural gas grid, introducing dimensionless parameters characterizing the gas and the solid sorbent feeds.

Results showed that the effect of the undesired CaO carbonation prevailed over the CaO hydration one; however, with a sub-stochiometric H_2 gas feed, an optimal condition was identified under which the outlet stream met the network specification. As for zeolites, due to the low adsorption capacity, the results suggested that only with a significantly higher amount of sorbent the SEM effect could be appreciable. These results highlighted the need for research on novel sorbents: selectivity towards water adsorption with respect to CO_2 appears to be crucial, suggesting that engineered zeolites might be preferable for this technology.

A Novel Approach for the Autonomous Control of Fluidized Beds with Intelligent Digital Twins Based on Transient Flowsheet Simulations

Monday, May 22, 2023 - 4:49pm to 4:51pm

Authors

Kräuter, R. - Presenter

Zhou, X., Hamburg University of Technology

Schiffner, S., Fraunhofer Institute for Factory Operation and Automation IFF

Pietsch-Braune, S., Hamburg University of Technology

Komarnicki, P., Fraunhofer Institute for Factory Operation and Automation IFF

Heinrich, S., Hamburg University of Technology

Heinrich, S., Hamburg University of Technology

Peglow, M., University of Magdeburg

Fluidized beds have been industrially established for decades. However, the process monitoring and control of such production plants are continuously confronted with new challenges to avoid undesirable process conditions and faulty production. Therefore, new control strategies which increase the efficiency of fluidized beds are of high importance. Such strategies can be provided by intelligent digital twins. A new framework for intelligent digital twins of fluidized beds, which is based on transient flowsheet simulations, is currently under development and is presented in this contribution.

The core of each digital twin is a so-called knowledge module which is a data processing unit for measurements and simulations. First, the knowledge module consolidates all online measurements of the sensors in the plant and detects possible outliers in the data. This data then serves as input for the transient flowsheet simulations. Second, a sensitivity study of the flowsheet simulations based on the input data and available control parameters is set up and is executed in parallel computation and significantly faster than real-time. Lastly, the simulation results are validated against the most recent measurement data. This procedure enables the knowledge module to accurately predict the future of the process and to calculate the necessary changes in the control parameters to ensure the desired process behavior.

The transient flowsheet simulations are carried out in the open source software Dyssol, which allows the implementation of any process and process chains in the field of solids process engineering [1]. For Instance, the data processing and consolidation of simulation results by the digital twin are shown for a lab-scale process of fluidized bed spray granulation.

[1] Skorych, V., Dosta, M., Heinrich, S.: Dyssol â€" an open-source flowsheet simulation framework for particulate materials, SoftwareX, Volume 12 (2020), 100572, DOI: 10.1016/j.softx.2020.100572

Development of an Artificial Neural Network EMMS Drag Model for the Simulation of Fluidized Beds in Chemical Looping Combustion

Authors

Stefanitsis, D. - Presenter, CERTH

Stamatopoulos, P.

Zeneli, M., National Technical University of Athens

Atsonios, K., CERTH

Nikolopoulos, N., Centre of Research and Technology Applications Chemical Process & Energy Resources Institute (CERTH/CPERI)

The current work presents an Artificial Neural Network Energy Minimization Multi-Scale (ANN-EMMS) drag scheme, which has been developed specifically for applications of fluidized bed reactors in chemical looping combustion. The data that feed the neural network are generated by a custom-built Matlab code that solves the EMMS equations for gas-particle conditions that correspond to those encountered in chemical looping combustion. The examined particles correspond to the various bed materials encountered in chemical looping combustion i.e., ilmenite ($FeTiO_3$), hematite (Fe_2O_3) and titanium oxide (TiO_2). The developed drag model is tested in a CFD simulation of a pilot scale circulating fluidized bed air reactor, which has been also performed as part of this work; the results of the pressure profile in the reactor were in good agreement with those of the conventional EMMS model.

Research on Non-Uniform Phenomena Due to Spatial Scale Effect in the Scaling up Process of CFB Boilers

Authors

Zhang, S. - Presenter

Yang, H. - Presenter, Key Laboratory for Thermal Science and Power Engineering of Ministry of Education, Department of Energy and Power Engineering, Tsinghua University

Yang, X., Key Laboratory for Thermal Science and Power Engineering of Ministry of Education, Department of Energy and Power Engineering, Tsinghua University

Zhou, T., Key Laboratory for Thermal Science and Power Engineering of Ministry of Education, Department of Energy and Power Engineering, Tsinghua University

Zhang, M., Key Laboratory for Thermal Science and Power Engineering of Ministry of Education, Department of Energy and Power Engineering, Tsinghua University

Circulating fluidized bed (CFB) boilers have the advantages of wide fuel adaptability, low-cost emission control, and width of load regulation range. So CFB boilers have been widely used as a clean coal combustion technology. To improve the efficiency of power generation, the scaling up process is the necessary development direction of CFB boilers. However, many non-uniform phenomena have occurred during the scaling up process caused by the effect of spatial scale, including the fluidization transformation of the gas-solid flow in the super high furnace, the gas-solid flow deviation under the condition of the multi-cyclone parallel arrangement, the vibration of external loops, the large margin and long-term pressure fluctuation of the large cross-section air distribution furnace, etc. These non-uniform phenomena seriously affect the stability and safety of the boiler operation. In recent years, combined with the actual running data of CFB boilers, targeted experiments have been designed and carried out. And corresponding physical models and mathematical models have been built and gradually improved. In this paper, these phenomena are introduced in detail based on the data from actual boilers and experiments. The important physical mechanism of the gas-solid flow behavior of CFB boilers is analyzed, including the multi-value of gas-solid flow under the multi-cyclone arrangement, the pulsation mechanism of the loop seal, the resonance mechanism of lateral fluctuations in the dense phase area of the furnace, etc. Eventually in this paper, effective engineering solutions are given to promote the scaling up process of CFB boilers and help the utilization of world energy and the development of the economy and society.

Influence of chemical phases on the mechanical strength of oxygen carrier bed particles

Authors

Purnomo, V. - Presenter

Mattisson, T., Chalmers University of Technology

Leion, H., Chalmers University of Technology

Faust, R., Chalmers University of Technology

Abdisa Ejjeta, L., Chalmers University of Technology

Sandell, E., Chalmers University of Technology

Hertzberg, M., Chalmers University of Technology

In a chemical looping setup, oxygen carriers bind oxygen from air through oxidation in an air reactor and releases it in a fuel reactor, oxidizing the fuel without the presence of nitrogen. Its ability to absorb and release oxygen in different conditions is the feature that makes chemical looping systems possible. Therefore, investigations of oxygen carrierâ€™s performance and characteristics are essential and necessary. Our study aims to examine the effect of oxidation states on physical properties of several iron-based oxygen carriers, such as their attrition tendency. Oxidation state of an oxygen carrier can affect its fluidization performance, especially in certain conditions that involve a more reduced oxygen carrier. The examined oxygen carriers in this work will be Norwegian ilmenite ore, synthetic ilmenite, Swedish copper slag, Swedish steel converter slag, and the Spanish Tierga iron ore. The oxygen carriers will be calcined and activated in a high temperature furnace and a fluidized bed batch reactor, respectively. The oxygen carrier will then be reduced to reach certain reduction levels, or mass conversion degrees, which correspond to different oxidation states. The attrition examination of these materials will be conducted in a customized jet cup rig, where the materials will be exposed to a harsh physical attrition at room temperature using air. Fine fragments from the materials will be trapped in a filter and weighed. Based on the weight data, the attrition tendency of the oxygen carriers can be compared and the effect of mass conversion degree of the oxygen carriers on it can be investigated. Furter investigations will be conducted with XRD, BET, and SEM/EDX. This will provide data on oxygen carriersâ€™ physical properties which are affected by the materialsâ€™ reduction degree.

A Novel Approach to Construct Equivalent Reactor Network (ERN) Models for a CFB Riser

Authors

Du, Y. - Presenter, China University of Petroleum, College of Chemical Engineering, Statekeylab of heavyoil processing

There always exists a conflict between model accuracy and computational speed for non-ideal reactors when modeling a chemical process. In view of this, a novel, easy to use, approach to construct compartment or equivalent ideal-reactor network (ERN) models is proposed for a circulating fluidized bed (CFB) riser based on its signal time-series. The pressure drop time-series is sampled during the Multi-Phase Particle-in-Cell (MP-PIC) simulation of the CFB riser. Detailed procedures for mapping the time-series to a directed weighted complex network (DWCN) and the determination of model parameters are presented. Effects of the sampling frequency and division scheme for the time-series on compartment/ERN models are also investigated. It is believed that the proposed methodology based on the time-series can be generalized to develop ERN models for non-ideal flows in the fluidization system. It is also expected that ERN models can be readily integrated into a process simulator, e.g., Aspen Plus, to resolve the conflict between model accuracy and computational speed in process simulation, on-line analysis and real-time optimization.

Today's Tools in the Scale up and Optimization of Today's Fluidized Processes

Authors

Parker, J. - Presenter, Oregon State University

Cocco, R., Particulate Solid Research, Inc. (PSRI)

Blaser, P.

Commercial fluidized beds have been around for a century with the development of the fluidized bed gasifier by Fritz Winkler. Historically, designing and scaling up these units was Edisonian because tools, concepts, and understanding were limited. Commercialization usually required several pilot scale units differing by conservative increases in size, yet the hydrodynamics were often poorly understood.

Today that is no longer the case. Cold flow modeling, probe development, and computational fluid dynamics (CFD) have reduced the number of pilot units needed, increased the scaling factors between iterations, and decreased the time to commercialization.

Today, the most effective methodology is to complement experiments with modeling in all stages of the scientific method. At the start, simple experiments provide vital inputs for the models. Well-defined models, in turn, are used to explore hypotheses and are valuable tools in building and managing the design space for additional and more sophisticated experiments. During and after the data collection, the models are further used to assist in the data analysis. What the models do not capture is equally important as what the models do capture, providing insights into the importance of phenomena beyond those used during model formulation. Key scale parameters are determined from both experimental and modeling results, along with invaluable engineer experience. Finally, the models once again become a vital tool in designing pilot, process demonstration, and full commercial-scale plants. The combination of these tools, used in the correct sequence, saves years of development efforts.

Multiple examples are presented, including fluidized catalytic cracking (FCC), gasification, and pyrolysis of waste plastic. Recent trends are explored whereby numerous sustainability applications are increasingly turning to fluidized processes.

The impact of todayâ€™s tools on the urgency of commercialization for todayâ€™s sustainability, renewables, and decarbonization applications is discussed, accelerating solutions for some of our most critical challenges.

Proof-of-Concept of a Novel Solid-Solid Heat Exchanger Based on a Double L-Valve Concept

Authors

Stefanitsis, D. - Presenter, The Centre for Research & Technology, Hellas

Zeneli, M., National Technical University of Athens

Papalexis, C., CERTH

Nikolopoulos, A., Centre of Research and Technology Applications Chemical Process & Energy Resources Institute (CERTH/CPERI)

Atsonios, K., CERTH

Nikolopoulos, N., Centre of Research and Technology Applications Chemical Process & Energy Resources Institute (CERTH/CPERI)

Indirect contact type heat exchangers (HEs) are broadly used in fluidized bed systems, usually in the form of heat transfer pipes containing a liquid or a gas medium. An alternative method of transferring heat to a gas-solid system is by beneficiating from the flow of another granular solid stream without necessarily requiring mixing of them. This can be mainly achieved though their continuous flow in concentric heat pipes. This paper deals with the proof-of-concept of a novel parallel-flow solid-solid HE, consisting of two L-Valves with concentric vertical tubes, named as Double L-Valve. This type of HE can be used in dual fluidized bed systems, as for instance in calcium looping process (oxy-fired or indirectly heated -IHCaL) for heating up the solid stream coming from the carbonator reactor by absorbing heat from the hotter stream coming from the calciner. This type of HE has been constructed on a small scale with acrylic transparent tubes and hydrodynamic tests have been conducted under cold flow conditions, in order to study the hydrodynamic response of the system at different designs and airflow rates. Apart from the cold flow pilot model, a theoretical assessment of the HE, operating under real conditions (hot flow conditions) at an industrial scale has been carried out by means of proper heat transfer models. Several parametric studies have been carried out to optimize its design, whereas its parametric costing has been undertaken. Finally, the structural integrity of the double L-Valve under high temperature conditions has been thoroughly investigated by Finite Element Analysis simulations and analytical models. All in all, the proposed HE concept offers several advantages to be integrated in a IHCaL system including design flexibility, simple construction, rather sufficient heat transfer efficiency, durability and rather low cost.

Experimental and Modelling Study of CO2 Adsorption for Carbonized Residues Using a Thermo-Gravimetric Fluidized Bed

Authors

Tondl, G. - Presenter, University of Natural Resources and Life Sciences, Vienna

Kronawitter, A., BOKU Vienna

Pfeifer, C., University of Natural Resources and Life Sciences

Higher value products from CO_2 and waste, can play a role in meeting climate goals as an application that contribute to a future net-zero CO_2 emission economy. The objective of the present work is to obtain KOH activated biochar from pyrolysed biogenic residue in a two-stage process and to evaluate its CO_2 adsorption performance in a thermo-gravimetric fluidized bed, operated as bubbling bed.

The fluidized bed TGA reactor consists of a 0.1 l reaction zone and is able to measure gas solid interactions such as drying, oxidation, adsorption and catalytic reactions during fluidization of the particles. It was validated with measurements of the adsorption isotherm of LEWATIT VP OC 1065 and the requested manufacturer data as well as results from literature data at a fluidization number of U/U_mf 2.05. The comparison with literature showed an R^2 of 0.929-0.0994 with respect to the conformity of the own collected data and the data from literature.

In addition, adsorption isotherms were determined for industrial carbon, non-activated biochar and activated biochar. A Langmuir-Freundlich model was applied to calculate the isometric heat of adsorption.

Improved adsorption of activated biochar compared to inactivated biochar was demonstrated. A maximum CO_2 uptake of 1.84 mmol/g was obtained at 50 °C and 100 % CO_2. A comprehensive comparison of all Langmuir models of all adsorbents show a comparable adsorption performance of the activated biochar with that of commercially produced activated carbon from hard coal.

The new developed fluidized bed reactor can help to develop cost reduced products from CO_2 and biogenic residues through multi stage processes adopted to locally available waste streams.

Pilot-Plant Investigation of High-Temperature Thermochemical Energy Storage Based on the Material System CaO/Ca(OH)2 in a Bubbling Fluidized Bed

Authors

Morgenstern, L. - Presenter, Technical University of Munich

Talebi, E., Technical University of Munich

Gleis, S., Technical University of Munich

Kerscher, F., Technical University of Munich

Spliethoff, H., Technical University Munich

Thermochemical energy storage based on the material system $CaO/Ca(OH)_2$ is currently investigated at the Chair of Energy Systems using fluidized bed technology aiming at storage temperatures of 400 °C to 600 °C for industrial applications. Besides many advantages such as high storage density, broad abundance, low costs, nontoxicity and chemical stability, one of the challenges to face is the mechanical particle stability. It restricts the number of overall storage cycles to be performed. Particle breakage with increasing number of storage cycles leads to a rising share of fines in the fluidized bed leading to defluidization eventually.

Previous studies in laboratory scale facilitated the understanding on storage material behavior in the fluidized bed. Hereby both, the hydration (discharging) and dehydration (charging) step are best performed in pure steam atmosphere close to equilibrium conditions. Charging and discharging reaction are controlled by temperature change. This knowledge is now transferred to a technical scale.

The reactor utilized here has a cylindrical fluidized bed volume of up to 30 L (overall volume of 100 L), a maximum operation temperature of 700 °C, a pressure rating up to 6 bar(g) and can be operated with pure air, nitrogen, steam or mixtures thereof. It is designed for particle collectives with d_{50} = 20 µm and higher.

First experimental results with calcination of the raw $CaCO_3$ (Sauter mean diameter is 219 µm) at 700 °C in pure steam atmosphere and subsequent cyclisation (successive charging and discharging) at a temperature difference of 50 K to the apparent reaction equilibrium are performed successfully. Results show promising scalability of the storage reaction in terms of material stability.

Investigation of Mixing/Segregation Patterns of Two Dissimilar Granular Solids in Fluidized BEDS By Capacitance Probes

Authors

Molignano, L. - Presenter

Troiano, M., Università degli Studi di Napoli Federico II

Solimene, R., Consiglio Nazionale delle Ricerche

Tebianian, S., IFP Energies nouvelles

Joly, J. F., IFPEN

Salatino, P., Università degli Studi di Napoli Federico II

Fluidized bed reactors are widely employed in industrial processes in which two or more type of solids are adopted. As example, in chemical looping combustion of solid fuels, the fuel reactor hosts the inert material, the solid fuel and the oxygen carriers. In most cases, solids present different features and operating conditions must be optimized to drive the fluidized bed reactor along the desired solids mixing status. A deep knowledge of the fluidization of dissimilar solids in size and density is required. As regards the experimental investigation of mixing/segregation of dissimilar solids in fluidized beds, the diagnostic tool for the measurement of solid concentration plays an important role.

In the present study, capacitance probes are adopted to assess the mixing between two dissimilar solids belonging to group B of Geldart classification, fluidized in the bubbling regime at 500°C. The technique enables the measurement of local mixture composition thanks to different dielectric properties of gas and solids and of the two chosen solids. However, the procedure to gather mixture composition in the emulsion phase of the dense bed at relevant operating conditions is not trivial. Indeed, the expansion state of the mixture dense phase is not known *a priori*, and its value is required to properly postprocess data from capacitance probes.

Different types of mixing and fluidization tests are carried out. Selected binary mixtures are characterized by analyzing capacitance probes signal during quenching tests, after fluidizing the bed for sufficient time. The void fraction adopted to postprocess the data is derived from loose beds bulk density measurements of mixtures of known composition. In a second type of experiment, capacitance probes signal is analyzed during the fluidization of the two solids. The voidage derives from blank characterization tests where single materials are fluidized. Finally, results from the two independent experiments are compared.

BIO-OIL Steam Gasification in a Fountain Confined Spouted BED

Authors

Lopez, G. - Presenter, University of the Basque Country

Cortazar Dueñas, M., University of the Basque Country

Alvarez, J., University of the Basque Country

Santamaria, L., University of the Basque Country

Fernandez, E., University of teh Basque Country

Garcia, I., University of the Basque Country

Amutio, M., University of the Basque Country

Olazar, M., University of the Basque Country

The combination of delocalized units for the fast pyrolysis of biomass to produce bio-oil followed by centralized units for the gasification of bio-oil appears as an economically attractive option for the full-scale production of syngas because transportation of bio-oil is less costly than that of biomass. This study deals with the use of the conical spouted bed reactor (CSBR) as a hybrid technology. Thus, bio-oil is first produced by fast pyrolysis under the conventional spouting regime and this liquid is then converted into syngas using steam as gasifying agent under the fountain enhanced spouting regime. This strategy is an interesting alternative to direct biomass gasification, as the cost-effective transport of bio-oil could make the implementation of this biomass valorization route profitable. Furthermore, this paper assesses the suitability of a bio-oil feeding device developed by our research group made up of of a line-thermostated at 60-80 ºC and a non-atomizing injector cooled by water. The bio-oil was produced at 500 ÂºC in the CSBR operating under conventional spouting, and collected in the condensation system. It was then gasified in the CSBR operating under fountain enhanced regime. The effect of temperature (800, 850 and 900 ÂºC) on the gasification performance was assessed. An increase in gasification temperature improved process efficiency in terms of syngas production and tar reduction, with the maximum carbon conversion efficiency being 96.3 % at 900 Â°C. Besides, the hydrogen content also increased with temperature from 4.51 wt% at 800 Â°C to 7.96 wt% at 900 Â°C on a dry basis, which resulted in an increase in H_2/CO ratio from 1.2 to 2, respectively, which are within the range required for GTL processes. Moreover, tar content decreased by around 70% in the 800â€"900 Â°C range, recording the lowest value (12.7 g/Nm^3) at 900 Â°C.

Calculation of the Nusselt Module in a Spouted Bed Reactor from Artificial Neural Networks

Authors

Saldarriaga, J. F. - Presenter, Universidad de los Andes

López, J., Institución Universitaria Colegio Mayor de Antioquia

The Nusselt modulus calculation is an important factor for the determination of the heat transfer of the particles. This calculation can be done by evaluating the fluid dynamics of the particles by means of the minimum fluidization velocity. On the other hand, artificial neural networks have been used since the last decade as a satisfactory alternative for the prediction of the fluid dynamic behavior of particles. Therefore, the objective of this work has been to develop a model based on artificial neural networks (ANN) suitable to quantify the influence of multiple factors on the rate of heat transfer in a conical bed reactor. The Nusselt module has been taken as the starting point and nine input factors have been evaluated, among which are the height of the bed, the diameter of the contactor, the angle of the cone and the minimum discharge velocity, among others. It has been found that the model fits appropriately to the equations proposed in the literature and can be used as a suitable model to predict the behavior of heat transfer in biomass-operated cone-bed reactors.

Regulating the Crude Oil to Chemical Process in a Multizone Fluidized Bed Reactor Using Intraparticle Heat Carriers

Authors

Zapater Bes, D. - Presenter, KAUST

Cui, M., King Abdullah University of Science and Technology (KAUST)

Dikhtiarenko, A., KAUST

Kulkarni, S., Ghent University

Shoinkhorova, T., KAUST

Aslani, I. A., KAUST

Alabdullah, M., KAUST

Mazumder, J., KAUST

Flores, R. I. M., KAUST

Alahmadi, A., KAUST

Alfilfil, L., Aramco

Morales Osorio, I., Aramco Research Center at KAUST

Almajnouni, K., University of Delaware

Gascon, J., Delft University of Technology

Castaño, P., KAUST

The catalytic cracking from crude oil to petrochemicals in one step is a profitable pathway for the future refinery. However, it requires significant improvements to solve some technological hurdles. Here we show a development of the catalyst formulation based on incorporating silicon carbide and its implications at different levels: particle shaping, thermal conductivity, fluidization and catalytic performance. The catalysts were prepared by spray drying using different catalyst formulations, SiC proportions, and particle sizes. First, we selected the best composition to shape our catalyst based on morphology and thermal conductivity. Then, we adopted the computational particle fluid dynamics method to systematically investigate the fluid dynamic and heat transfer characteristics of the selected catalyst compared with a reference commercial one in a multizone fluidized bed reactor. Finally, we compare the catalytic performances of the two catalysts during the crude oil to chemical process. Our results show that with careful consideration of all these catalytic and hydrodynamic parameters, we improve the overall process performance at both the catalytic and reactor levels with higher light olefin yield of 22%.

Measurement of Bubble Property Distributions for Better Description of Mass Transfer Limitations in Chemical Reactors

Authors

Riechmann, P. - Presenter, Paul Scherrer Institut

Schildhauer, T., Paul Scherrer Institut

Tomographic X-ray measurements of fluidized beds were evaluated by pseudo three-dimensional reconstruction of individual bubbles in a freely bubbling fluidized bed. It was found that several aspects of the applied methodology significantly influenced the results: i) the consideration of the individual rise velocities of the bubbles for the calculation of the vertical dimension of each bubble, ii) the determination of the surface area and shape of each bubble which turned out to be correlated with the rise velocity, as well as iii) a new adaptive threshold method for the binarization of the reconstructed void fractions, ensuring that the measured void fraction is maintained throughout the reconstruction procedure.

The X-ray tomography measurements showed that the bubble properties are broadly distributed for a given location within the investigated BFBs. This indicates that the description of the hydrodynamics in BFB merely by mean values of the bubble properties might not be sufficient. This is especially relevant for the prediction of critical reactor characteristics such as mass transfer limitations which could be caused by bubbles that are faster and larger than the corresponding mean values suggest.

Another important findings are the strong correlations between the size, the rise velocity and the shape of the bubbles. These correlations are particularly important in the context of measurements with optical probes or similar methods.

Correlations from literature that predict one average bubble size as a function of the fluidization number and measurement height did not describe the data well, especially for higher fluidization numbers.

The individual bubble property distributions can be described with generalized extreme value distribution (GEV) functions and their correlations can be accounted for with t-copula functions. Data sampled by these distribution functions can be used for the verification of CFD models or as a foundation for the development of semi-empirical correlations.

Simulation of a Granular Gas with Non-Spherical Particles - A Stochastic Modeling Approach

Authors

Gbolasere, M. - Presenter, Purdue University

Morris, A., Purdue University

Many industrial systems that involve particle technology utilize media that contain irregularly shaped particles. However, modeling the fluid dynamics of non-spherical particles (NSPs) is challenging due to the complex nature of the inter-particle interactions. The discrete element method (DEM) can be used to model non-spherical particle flows, but the computational cost associated with contact detection limits the application of DEM to small systems. We develop a Monte Carlo methodology where simulators are used to represented a large number of real particles. Data from high-fidelity deterministic models are used to generate probability distribution functions (PDFs) that predict post-collisional states of particles without the need for costly contact detection algorithms. Such post-collisional state distribution functions capture scattering, translational-rotational energy exchange, and orientation. These PDFs are then integrated into the Monte Carlo framework and the new method predictions are compared to DEM simulations for different particle shapes. Preliminary results of the evolution of rotational energy in a homogeneous system of spherocylinders (A.R = 1.5) versus mean collision time, and a close match between the DEM and DSMC is observed. Due to the elongated nature of high aspect ratio particles, a significant amount of energy is stored in the rotational modes, and the particle shape thus affects how the system cools and clusters over time. Correlations for the cooling rate for different particle shapes are presented.

Sink and/or float in gas-fluidized beds

Authors

Tsuji, T. - Presenter, Osaka University

Sinking/floating motions of large solids such as biomass fuel pellets and incombustible wastes in gas-fluidized beds are important in several applications. It has been discussed based on the balance between the buoyant force due to the gas pressure gradient in the beds and the gravity force, characterized by the ratio between the material density of the large object and the bulk density of the bed material (r_{object}/r_{bed}). Although this simple description works properly as a first approximation, we observed some deviations from it simultaneously; non-uniformities of the beds make their behaviors complex. This study explores the details of sinking/floating motions using the numerical model we developed for dense gas-solid flows with large size differences based on the DEM-CFD coupling technique. With the help of collaborators, we performed experiments first and then performed the corresponding numerical simulations. We found non-trivial dependencies on gas velocity, object shape, and local concentration; some objects expected to sink float and vice versa. In highly-concentrated cases, we found that some objects having the same mass and volume float, and some others sink. These behaviors cannot be characterized only by the simple averaged picture and the importance of local flow fields around the objects is emphasized. The physical mechanisms based on the local flow fields will be revealed in the presentation.

Comparison of Pumping Costs in Different Fluidization Technologies for CSP Applications with SiC Particles

Authors

Corcoles Tendero, J. I. - Presenter

Domínguez, P. - Presenter

Díaz-Heras, M., Castilla - La Mancha University

Almendros Ibáñez, J. A., Universidad de Castilla-La Mancha

The feasibility of using fluidized beds for thermal energy storage systems with direct radiation on the particles has been sufficiently demonstrated in several previous works. However, the focus has been placed on the useful energy stored, leaving the costs or energy losses derived from pumping particles in the background.

This work proposes to compare the pumping powers using the same material in different configurations: spout-fluid, spouted and bubbling beds. Within the bubbling fluidization technology, two different air inlet distributions have been applied to the bed: even or homogeneous and uneven fluidization. In the same way, spout-fluid and spouted present the nuance of the cylindrical bed with different air inlets (only central part or air distribution through the center and periphery).

The aim of this work is to highlight the necessary costs of fluidizing the same material in different fluidization technologies, quantifying in terms of electrical power the air necessary to carry out the fluidization of the particles. The material used in this work is SiC, typically used in Concentrated Solar Power applications, with a particle diameter of 0.56mm. Its characteristics, minimum fluidization (U_{mf}=0.24m/s) and spouting (U_{ms}) velocities were previously measured: depending of height and mass: U_{ms}=0.76 m/s (h=12cm), U_{ms}=0.83m/s (h=15cm), 0.95m/s (h=17cm).

The most advantageous bubbling bed air distribution is considered to achieve similar results to even fluidization but with a 26.4% reduction in pumping costs. On the other hand, the conical spouted bed presents a 52.3% reduction in pumping costs compared to the spout-fluid bed for the same material. It should be noted that the pressure drop of the mass of particles has a notable contribution to the estimated total electrical power.

Numerical Study of the Fluidization of the Binary Types of Particles Using SOM-KTGF-MT

Authors

Sun, D. - Presenter, National Institute of Clean and Low Carbon Energy

The second-order moment of kinetic theory of granular flow (SOM-KTGF) has been verified in predicting the particle-laden flow in the median concentration, by the four-way coupled second-order method (FW-SOM) established via coupling the gas phase second-order turbulence model [1]. The anisotropic SOM-KTGF remedied the shortage of the classic isotropic KTGF in predicting the â€œdiluteâ€� phase of particles in the fluidized risers. These particles of the dilute phase in the fluidized bed were suspended in the reactor and less affected by the other particles surrounded. As a result, these particles showed the behaviour of the high inertia and high anisotropic fluctuation. In the previous study, the FW-SOM and SOM-KTGF demonstrated the superior to the classic KTGF in the numerical study of the fluidization.

A novel model of the SOM-KTGF for multi-type particles (SOM-KTGF-MT) was proposed in this study to investigate the fluidization of the particle mixture with binary types. The particles in the mixture differ in all types of the physical properties such as the density, size, and restitution coefficient et al. In the model, SOM-KTGF-MT, the conservations equations of the volume fraction, velocity and the velocity fluctuation of particles were solved for each type of particles. In this study, SOM-KTGF-MT was applied to simulate the flow behaviour of gas-solid flow with different size and density of particles in the fluidized bed. The anisotropic fluctuation of the binary types of particles and effects of the particle-particle interaction were investigated in terms of the particle distribution and turbulent characteristics.

[1] D. Sun. Four-way coupled second-order method of two-fluid model for gas-particle flow and the numerical simulation in horizontal channels. Powder Technology 390 (2021) 354â€“368

Fluidization of Size Segregating Binary Mixture

Authors

Gamit, H. G. - Presenter

Bose, M., Indian Institute of Technology Bombay

Seethamraju, S., Indian Institute of Technology Bombay

Singh, N., Indian Institute of Technology Bombay

Fluidization of a mixture of bi-disperse particles is often characterized by initial and final fluidization velocities. Both conditions are typically determined following the conventional method based on the behaviour of the bed pressure drop vs the superficial gas velocity. In early days, manometers were used for measuring bed pressure drop. Later with piezoelectric devices, the fluctuations in the pressure-drop in a fluidized bed is more accurately captured. In addition, visualization of the bed, albeit peripheral, helps in understanding the physical structure of the bed. The fluctuations in the bed pressure drop vs the superficial gas velocity plot and the physical configuration of the bed suggest possible existence of multiple characteristic points in a fluidized bed of mixture. The objective of the current work is to understand the fluidization and de-fluidization process of a binary mixture using simultaneous measurement of pressure drop and video imaging of the bed.

Particles of two different mean diameters, both from Geldart B group, are selected for experiment. Particles of different diameters are coloured differently for clear demarcation.

Experiments were performed in a transparent glass column for visualization. Videos of the bed were captured using a DSLR (Nikon D5200) camera. A pressure transmitter (AFRISO DMU11D) was used to measure the pressure drop. Data were collected using the National Instrumentâ€™s data acquisition system (NI 9219). The extracted images from the videos were analysed using `ImageJâ€™ software.

Experiments were performed for different bed configurations. Intermediate configurations and characteristic points depend on the initial condition of the bed. Different phenomena such as vertical and azimuth segregation and square slug formation are observed depending on the initial condition; however, the final configuration after the complete fluidization - de-fluidization process is independent of the initial bed condition.

Simulation of Hydrodynamics in a Gas-Solid Downer Reactor

Authors

Aldugman, T. A. - Presenter, King Abdullah University of Science and Technology

Cui, M., King Abdullah University of Science and Technology (KAUST)

Alzailaie, A., King Abdullah University of Science and Technology

Alhareth, A., King Abdullah University of Science and Technology

Langley, K., King Abdullah University of Science and Technology

Alfilfil, L., Aramco

Morales Osorio, I., Aramco Research Center at KAUST

Almajnouni, K., University of Delaware

Thoroddsen, S., King Abdullah University of Science and Technology

Castaño, P., KAUST

The downer reactor is a promising technology that could potentially replace the current riser reactor technology used in fluid catalytic cracking (FCC). The main benefit of the downer is the very narrow distribution of contact times between the solid (catalyst) and the gas (reactive phase), which can lead to improved selectivity control. This work aims to compare the hydrodynamic performance of the two modes of operation of the downer reactor (co- and counter-current) in a pilot plant scale unit: 1.75 m height and 27 mm ID. To achieve this goal, we performed a series of experiments in a downer reactor at ambient conditions, using a commercial FCC catalyst and measuring the pressure drop over various conditions. Then we coded a simulation using a computational particle fluid dynamics (CPFD) approach and validated it by the previously performed experiments. Finally, we used the validated simulation to explore the hydrodynamic performance of each mode of operation and compare one against the other. The simulations show that the counter-current downer reactor has a 69% higher solid holdup than the co-current one, 98% longer residence times of the solid, and both modes approach the plug-flow behavior (for the solid) compared to the riser reactor.

Effect of Particle Size on Solids Distribution in Dense Gas-Liquid-Solid Flow in a Slurry Bubble Column

Authors

Manna, S. - Presenter

Buwa, V., Indian Institute of Technology Delhi

Slurry bubble column reactors are predominantly used to perform solid-catalysed gas-liquid reactions such as resid-hydrocracking, etc. In addition to other operating and design parameters, particle size distribution (PSD) of solids can have a significant influence solids distribution and consequently on the performance of slurry bubble column (SBC). In most of the previous studies, either mono-dispersed particles or particles with narrow size distribution are considered. Mixing and segregation of particles of different size and their influence on solids distribution and hydrodynamics are not yet well understood, which is the main objective of the present work. In this work, using Eulerian multifluid CFD model and KTGF, we have investigated the effect of PSD on mixing and segregation of particles of different size and their influence and solids distribution and hydrodynamics.

We have implemented widely-used closure models for gas-liquid and liquid-solid phase interactions in the commercial solver and performed simulations of air-water-glassbead system with a constant particle size (d_p=250 Î¼m), for wide range of superficial gas velocities (U_g=5-30 cm/s) covering homogeneous and churn turbulent flow. We compared the predictions of overall gas volume fraction and gas volume fraction distribution measured earlier in our research group. The predictions were further improved by incorporating non-drag inter-phase coupling forces along with bubble swarm corrections. We have used the experimentally-validated model to understand the effects of PSD on solids distribution, segregation and mixing of particles, for a wide range of U_g (=5-30 cm/s). In particularly, we are working on understanding the mechanisms of segregation and mixing in bubbly and churn turbulent flow conditions. Further details of the computational model, experimental validation and result will be presented at the conference. Eulerian CFD models, capable of accurately predicting the solids distribution and segregation/mixing characteristics will be very useful in design and performance optimization of three-phase slurry reactors.

Investigation of Particle Properties and Dynamic Modelling of Fluidized Bed Spray Granulation for Its Autonomous Control with Intelligent Digital Twin

Authors

Zhou, X. - Presenter, Hamburg University of Technology

Kräuter, R.

Pietsch-Braune, S., Hamburg University of Technology

Heinrich, S., Hamburg University of Technology

During fluidized bed spray granulation process, the particle moisture content and bed temperature are of high importance due to their influence on the final product properties such as morphology, flowability and hardness. However, the classical time-consuming loss on drying method for measuring particle moisture content usually leads to delayed process control and defect production. Hence, inline monitoring of particle moisture content becomes necessary. Additionally, the influence of process parameters on the particle moisture content and bed temperature should be modeled with flowsheet simulation to ensure a faster process prediction, which contributes to the development of a digital twin for the autonomous control of fluidized bed processes.

In this work, microcrystalline cellulose particles are coated in a lab-scale fluidized bed spray granulator, with the inline measurement of the particle moisture content based on microwave transmission. Important process parameters are varied, including the flow rate and the temperature of the fluidization air, the water spray rate and the bed mass. The resulting inline particle moisture content and bed temperature are correlated with the process parameters using the response surface methodology. The desired process parameter ranges are determined.

The dynamic behavior of particle moisture content and bed temperature during granulation are modeled based on drying kinetics. The established models are implemented in the open-source dynamic flowsheet simulation software Dyssol [1] and validated with experimental results. The simulation input parameters and results serve further for the development of a digital twin for the autonomous control of fluidized bed spray granulation.

[1] Skorych, V., Dosta, M., Heinrich, S.: Dyssol â€“ an open-source flowsheet simulation framework for particulate materials, SoftwareX, Volume 12 (2020), 100572, DOI: 10.1016/j.softx.2020.100572

Bubble Dynamics in Gas Fluidisation of Nonspherical Particles

Authors

Zhou, Z. - Presenter, Monash University

Shrestha, S., Monash University

Yu, A., Monash University

Bubbling fluidized beds (BFB) have widespread applications in various industrial process because they have unique features such as high chemical conversion, heat and mass transfer and mixing. Bubbles generated are generally regarded as the driving force influencing the performance and efficiency of BFB. Several variables can affect the bubble dynamics such as fluidization gas velocity, particle shape, inter-particle force, etc. Understanding the effect of these variables is of paramount importance to the improvement of design, operation and control of BFB. In this presentation, the aim is to report how the particle shape affects bubble dynamics under via numerical simulation under different conditions. Ellipsoidal particles are used as they have the advantage in representing a large range of particle shapes varying from disc-like to rod-like. The results reveal that for the case of continuous single jet, particle shape can alter the mechanisms of bubble behavior such as splitting and coalescence. Ellipsoids have larger bubble equivalent diameters, more irregular bubble shape, and lower bubble frequency and bubble velocities. For the case of uniform BFB, the results show that bubble flow patterns for ellipsoids are asymmetric leading to different solid flow pattern, solid mass flux, and mixing characteristics in the whole bed. The ellipsoids have a smaller bubble size and lower bubble rising velocity than spheres. The combined effects of particle shape and van der Waals force on bubble dynamics are also examined. It is found that the bubble coalescence and splitting phenomena are suppressed with the increase of van der Waals force. The bubble diameter and velocity decrease with the increase in extent of van der Waals force for particle of different shapes. Moreover, the oblate/prolate spheroids transform to non-bubbling fluidization under the influence of high cohesive force while spheres form channels.

Dynamic Bubble Tracking in Fluidized Beds Via Electrical Capacitance Volume Tomography

Authors

Lindmüller, L. - Presenter, Hamburg University of Technology

Watson, B., Hamburg University of Technology

Heinrich, S., Hamburg University of Technology

Theuerkauf, J., The Dow Chemical Company

Yao, Y., Engineering and Process Science, Core R&D, The Dow Chemical Company

Fan, Y., The Dow Chemical Co

Capecelatro, J., Dept of Mechanical Engineering

Detailed knowledge of the dimensions and velocities of bubbles in fluidized beds is invaluable for the development and validation of correlations and models of particles in fluidized bed systems. This work sets forth the development of a novel, dynamic bubble detection algorithm for use with electrical capacitance volume tomography (ECVT) sensors for bubbling fluidized beds. The novel phase detection method was developed to address the shortcomings of other published methods by more fully utilizing the three-dimensional capability of the sensors. After the sensor parameters were optimized to verify capability of bubble detection, measurements were made with glass beads as well as quartz sand. A range of air velocities in a 10 cm diameter bubbling fluidized bed were used. The results largely agree with the fundamentals of bubbling fluidization and results from publications with similar experimental setups. Within each measurement the bubble dimensions, velocities and frequencies over the axial and radial position could be evaluated. Due to the three-dimensional nature of the novel bubble detection technique, insight into the directional tendencies of detected bubbles was gained. For example, bubble migration toward the radial center of the bed, as well as radial and axial bubble coalescence and splitting, are more clearly evident from the trends that are producible with this method.

A Quantitative Relationship between the Magnitude of Interparticle Forces and the Hausner Ratio

Authors

Chaouki, J. - Presenter, Ecole Polytechnique Montreal

Soleimani, I., Polytechnique Montreal

Shabanian, J., Polytechnique Montreal

Interparticle forces (IPFs) significantly affect the performance of gas-solid fluidized beds with Geldart groups A and C powders. Different types of IPFs can be present in a gas-solid fluidized bed, including van der Waals, electrostatic, liquid bridge, and solid bridge or sintering forces. Models have been presented in literature to fundamentally describe and estimate these types of IPFs for micron-sized particles. A critical challenge in applying these models to bulk particulate systems is their dependence on many factors for an adequate estimation of the level of IPFs. These parameters include particle diameter and shape, surface properties, conductivity, Hamaker constant, operating conditions, humidity, etc. The measurement of some parameters in these models, e.g., the curvature of liquid bridge, are hardly possible and others, e.g., the surface roughness for fine irregular-shaped particles, are hardly reliable. Hausner ratio (HR) is defined as the ratio of tapped to loose bulk densities. It has long been employed for a qualitative evaluation of IPFs, i.e., a higher HR is generally translated to a higher degree of cohesivity. However, no equation has been proposed to relate these two quantitatively.

A simple model is presented in this work to describe the variation of the bed voidage and HR with the size of agglomerates for Geldart group C powders. The model suggests that the magnitude of IPFs, i.e., the ratio of IPFs and weight of a particle, is inversely proportional to the fourth power of solid phase fraction of the tapped bed. The proposed equation can quantify the magnitude of resultant IPFs acting on each particle irrespective of the type of IPFs under various conditions, including ambient and elevated operating conditions. The quantified IPFs by the proposed equation for several powders agreed with the results of other experimental approaches for estimating these forces.

Use of a Vibrofluidizing Bed for Drying Orange Bagasse

Authors

Saldarriaga, J. F. - Presenter, Universidad de los Andes

Barbosa, A. M., Universidad de los Andes

Rocha, T., Federal University of San Carlos

Freire, J. T., Federal university of São Carlos

Freire, F., Federal University of San Carlos

The vibrofluidized bed is a promising technology for drying irregular materials and high humidity, which in another type of technology would be complex due to the clogging produced. Brazil is the largest orange producer in the world, part of the production is directed to the production of industrialized juice that generates orange bagasse as waste. The objective of this work was to evaluate the influence of the pretreatment with virgin and hydrated CaO in the mechanical dehydration stage and in the thermal drying in the vibrofluidized bed. In the mechanical dehydration stage, a moisture reduction of 1 to 0.6 (dimensionless) was obtained with the treatment with CaO. No significant variations in the type of CaO used in this step were identified, however, the increase in virgin CaO content caused an increase in flux. Finally, the kinetics of drying in a vibrofluidized bed with amplitude and vibration frequency of 0.015m and 6.16 Hz (Î“ = 2.29), respectively, at 70 and 80Âº C, where the humidity was reduced to approximately 0.1, was analyzed. The drying kinetics compared the influence of the CaO pretreatment under different conditions and, in general, the use of CaO improved the drying process. Due to the characterization of some other dry products, there is a high HHV, in addition to containing a high volatile matter. The latter allow good combustion conditions, obtaining renewable energy and valorizing waste.

Bed-Wall Frictional Loss in a Bubbling Bed with Horizontal Flow of Solids

Authors

Farha, M. - Presenter, Chalmers University of Technology

Guío-Pérez, D. C., Universidad Nacional de Colombia

Isaac Nimvari, M.

Pallarès, D., Chalmers University of Technology

Johnsson, F., Chalmers University of Technology

The establishment of a horizontal fluidized solids flow is a feature present in dual-fluidized bed systems with at least one reactor operated under bubbling conditions. Frictional loss between horizontally-flowing bed solids and reactor walls is a crucial aspect of reactor design since this loss limits the horizontal flow rate within the system.

The present study aims to develop a description of the bed-wall friction for bubbling beds with a horizontal flow of solids. This work uses experiments under ambient fluid-dynamically down-scaled conditions since these provide good operational flexibility and the possibility of implementing measurement techniques for detailed flow monitoring while allowing scale-up of the results to large-scale hot conditions. Bed material of Geldart-B type with a particle density of 8770 kg/m^3 and a mean size of 125 µm was used. The fluidized bed used is operated under bubbling conditions and consists of a closed horizontal loop configured to force a horizontal circulation of solids which can be controlled. For this purpose, the horizontal loop consists of two sections that are operated independently: a relatively small conveying section (forcing the solids to flow horizontally) and a transport section (through which the horizontal flow of solids is returned to the conveying zone). The transport section has an average length of 1.12 m, which scales up to 9.35 m at 850 â°C conditions. The system is equipped with in-bed pressure probes to sample pressure drop along segments of the transport section, which is assumed to represent the frictional pressure drop. Additionally, the horizontal solids velocity established was measured through the magnetic solids tracing technique. Variations of conveyed solids flow (u_0/u_{mf} within 3â€"100), channel width (0.04â€"0.12 m, scaling up to 0.33â€"1 m), and bed height (0.08â€"0.1 m, scaling up to 0.67â€"0.83 m) are used to parametrize the results.

Investigation on Conveying Mechanisms for Forcing Solids Horizontal Circulation Under Bubbling Fluidized Bed Conditions

Authors

Farha, M. - Presenter, Chalmers University of Technology

Guío-Pérez, D. C., Universidad Nacional de Colombia

Johnsson, F., Chalmers University of Technology

Pallarès, D., Chalmers University of Technology

The solids circulation rate is one key fluid-dynamic parameter of dual fluidized bed (DFB) systems, as it profoundly impacts the mass and energy transport between reactors and, thus, the overall system performance. Therefore, efficiently forcing a controllable flow of solids is a strongly desired feature in the design of DFB systems. The solids circulation in DFB systems with circulating fluidized beds (CFB) is typically attained by the solids elutriation from the riser. In contrast, DFB systems with interconnected bubbling beds lack knowledge of how efficiently fluidized solids can be transported between the reactors.

This work investigates different solids conveying mechanisms to induce a controlled horizontal circulation of fluidized solids at bubbling conditions. A closed horizontal loop configuration was used in the present study to circulate Geldart B solids in a bubbling fluidized bed. The experimental setup comprises a conveying section inducing the solids circulation and a transport section feeding back the bubbling fluidized solids to the conveying section. The design and operation of the unit follow the Glicksman scaling laws, and the closed loop with a characteristic length of 1.24 m operated at ambient conditions scales up to 10.35 m at hot conditions typical for solid fuel conversion (gasification, in this case). The resulting solids horizontal circulation rate is measured using magnetic solids tracing. Furthermore, the differential bed pressure induced in the conveying section is monitored, which allows the calculation of an energy-based efficiency of the conveying zone.

The experimental test matrix contains variations of bed height, gas flow rate provided to the solids conveying zone, and geometric configuration of the conveying zone. Regarding the latter, five different configurations were tested, each exploiting distinct features of the gas-solids flow to induce a horizontal solids flow: directed gas injection, free solids splashing, confined solids splashing, slugging, and solids entrainment.

Particle Circulation and Mixing Kinetics in Bubbling Fluidised Beds Using Pept

Authors

Windows-Yule, K. - Presenter, University of Birmingham

Werner, D., University of Birmingham

Wellings, A., Recycling Technologies

Seville, J., The University of Birmingham

Thorpe, R. B., School of Engineering

Bhattacharya, S., Recycling Technologies

Sanchez, A., Recycling Technologies

Besong, M., Recycling Technologies

Prediction of the solids motion and the kinetics and extent of mixing in fluidised beds is important in a new generation of industrial processes including, for example, thermal treatment of mixed plastic waste as part of the circular economy in plastics. Verification of classical two-phase type predictions at industrial scale has been limited by lack of appropriate non-invasive measurement approaches. We describe the use of a new generation of Positron Emission Particle Tracking (PEPT) software and hardware [1] to track particle motion in scaled laboratory and pilot-plant fluidised beds up to 1.5m in diameter, giving more detailed solids motion information than ever before obtained, including complete motion maps at 100Hz acquisition rates and ~1mm accuracy. Time averaged bed circulation patterns generally show the behaviour predicted by Geldart, with strong upward motion towards the centre-line and down at the walls. However, PEPT trajectory analysis at all scales also shows the discontinuous â€œjumpâ€� and â€œidle timeâ€� behaviour observed earlier by Stein *et al.* [2]. This is significant for reactions and in processes such as agglomeration, which depend on local particle contact. The measurements obtained also enable indirect measurement of bubble velocity. As expected, mixing time is related to individual particle circulation time. We demonstrate the use of the trajectory analysis method first proposed by Martin *et al.* [3] for measurement of dispersion and mixing rate, showing the dependence of mixing time on excess gas velocity and the relationship to particle circulation time. We demonstrate the use of the Barracuda computational software to predict particle motion in equivalent circumstances.

[1] Windows-Yule *et al* 2022 *Rep. Prog. Phys.* **85** 016101

[2] Stein *et al* 2000 *Chem. Eng. Sci.* **55** 5291

[3] Martin *et al* 2007 *Chem. Eng. Sci.* **62** 3419

Characterization of the Solids Flow in Circulating Fluidized Beds By High-Frequency-Radar Measurements

Authors

Guío-Pérez, D. C. - Presenter, Universidad Nacional de Colombia

Bonmann, M., Chalmers University of Technology

Johnsson, F., Chalmers University of Technology

Pallarès, D., Chalmers University of Technology

This work presents results from the latest developments in the use of the radar technology for the diagnosis of solid particles flow in fluidized beds. We apply a radar system at submillimeter wave-frequencies (0.34 THz) operated in the frequency-modulated continuous-wave pulse-Doppler mode, which was previously shown to be able to measure concentration and velocity of small solid particles. The radar system used provides a time resolved (up to 30 Hz) measurement along the beam (4â€“6 cm wide and 0.5 cm longitudinal steps, yielding a spatial resolution of 6.3â€“14 cmÂ³). This work further explores the capabilities of the radar technology with the aims of both expanding the validity range of the technique and examining the solids flow in specific regions of a circulating fluidized bed. With this purpose, measurements along the vertical and horizontal directions in riser (3.1 m high and 0.45 mÂ² in cross section, using 106 Âµm glass beads as bed material) were performed, for which fluidization velocity and total solids inventory were varied to achieve a variety of flow conditions.

We report the following achievements:

- A more accurate evaluation than the previous work with respect to the agreement between the solids concentration values measured by means of radar and pressure, especially for concentration levels $>1\cdot10^{-1}$ m^3/m^3 for which the penetration of the radar is shortened. This was attained by performing radar measurements along the horizontal direction in the riser.
- Confirming the validity of the radar technique for measurement of local concentrations by combining measurements performed in the vertical and horizontal directions.
- Calculation of solids mass fluxes in both horizontal and vertical directions at specific locations in the bed by combining the measurements of solids concentration and distribution of solids velocity.
- Evaluating the solids concentration, velocity and flux profiles near the vertical walls and close to the riser exit.

Modelling of Gas-Solid-Liquid Flow and Particle Mixing in a Rotary Drum

Authors

Tang, X. - Presenter

Yue, Y., UNSW

Wang, S., UNSW

Shen, Y., UNSW

Solid-liquid rotary drums have been widely practised in various industries,while the complex multiphase hydrodynamics hinders the understanding and optimization of these processes.in this work,the computational fluid dynamics-discrete element method coupled with a volume of fluid is developed to describe the gas-solid-liquid flow and mixing in a rotary drum including inter-particle collisions,inter-phase interactions and interface morphology. A smoothing method is used to link the quantities between the particle and computational grids,allowing the fine grids to resolve flow details such as the interface and curvature.The effects of liquid presence and rotating speed on particle-scale behaviors (e.g.,repose angle,active-passive zone,solid residence time and force chain) and the mixing performances such as mixing index and dispersion are studied.The results show a positive correlation of the active depth,mixing degree and particle dispersion with the rotating speed.The liquid presence leads to a deeper active depth,prolonged solid residence time in the active zone,and lower contact force.Thw work sheds light on the design and process optimization of rotary drums.

Spout Geometry of Fine Particle Conical Spouted Beds Equipped with Fountain Confiner

Authors

Bolanos, M. - Presenter, University of the Basque Country (UPV/EHU)

Sukunza, X.

Estiati, I., University of the Basque Country

Tellabide, M., University of the Basque Country

Altzibar, H., University of the Basque Country

Arabiourrutia, M., University of the Basque Country (UPV/EHU)

Olazar, M., University of the Basque Country

The spouted bed regime is an alternative gas-solid contact method to fixed and fluidized beds for the treatment of coarse particles (d_p >1 mm). Despite its benefits, the main limitation of this technology lies in its scaling up. Accordingly, the ratio of the gas inlet diameter to particle diameter, D_0/d_p, cannot be higher than 20-30. The use of internal devices, such as draft tubes and fountain confiner, have allowed to increase the D_0/d_p ratio to 1000, and ensure stable operation even with ultrafine particles (d_p<0.1 mm).

With the aim of designing spouted beds for industrial applications, heat and mass transfer rates must be optimized. These two parameters are significantly conditioned by the cross-flow of solid particles from the annulus into the spout, which depends on the spout geometry. Accordingly, runs have been carried out in a fountain confined conical spouted bed to assess the spout-annulus interface along the bed, which is the information required to determine the average spout diameter. With this aim, an optical system fitted with a borescope has been used to track particles in the bed. These runs have been carried out in a wide range of contactor angles, air inlet diameters and draft tubes differing in the aperture ratio. The effect of the air inlet velocity on the spout diameter has also been ascertained.

The results show that the spout expansion increases when the air velocity is increased. However, the effect of the contactor angle, the draft tube aperture ratio and the air inlet diameter is not straightforward, as their individual effect depends on the other geometric factors. Regarding the spout geometry, the maximum expansion is observed close to the bed surface. The results obtained allow a better understanding of fine particle spouting hydrodynamics in conical beds, which is essential information for their scaling up.

A Method for Direct Measurement of Electrostatic Charge Generated By Particle-Wall Friction in Gas-Solid Fluidized Beds

Authors

Kim, S. - Presenter, Korea national university of transportation

KIM, S. W., Korea National University of Transportation

Electrostatic charges generated by continuous friction between particles and between particles and the reactor wall in gas-solid fluidized beds can cause problems such as unstable operation and reduced productivity and product quality. Various measurement studies have been conducted to identify factors that affect electrostatic charge generation in fluidized beds. However, research on direct measurement methods for electrostatic charge generation on the metal surface of the reactor is lacking. In this study, we proposed a electrostatic charge measurement technique near the metal surface of the reactor wall.

Experiments were conducted in a cold flow model of a gas-solid fluidized bed reactor (50.8mm-i.d. x 600mm-high) with polyethylene particles that generated significant electrostatic charges. An inner sensing electrode (50.8mm-i.d. x 20mm-high) made of SUS material that does not interfere with gas-solid flow was installed to measure electrostatic charges under various fluidization conditions.

As the gas velocity and fluidization time increased, the amount of electrostatic charge accumulation within the particles increased due to increased particle-particle contact and particle-wall friction. The electrostatic charge by the developed probe showed opposite polarity results compared to the external measurement method commonly used in literature and demonstrated high sensitivity reflecting the direct behavior of the particles.

Thermocatalytic Decomposition of Methane in a Fluidized Bed Reactor: Proof of Concept

Authors

Hadian, M. - Presenter

Buist, K., Eindhoven University of Technology

Bos, R.

Reesink, B., BASF Nederland bv

Kuipers, J. A. M., Eindhoven University of Technology

Thermocatalytic decomposition of methane (TCD) into functional-carbon and hydrogen requires optimal design of the reactor and the process. To facilitate this, the particle growth needs to be characterized and understood by means of conducting kinetic measurements and modeling of the heat and mass transfer and catalytic reaction at the particle and reactor scale.

TCD was studied in a lab-scale fluidized bed reactor to assess the effect of reactor parameters (such as temperature, concentration, particle size and space velocity) and investigate the effect of the increasing particles size and mass during reaction on the fluidization and finally to derive a kinetic model for the reaction and deactivation of the catalyst. A commercial nickel catalyst supported on silica was used in our kinetic experiments. The performance of the catalyst was evaluated and values up to and higher than 70 g of functional Carbon NanoFibers (CNFs) per gram of catalyst was produced. The fluidization regime was changed from a turbulent regime with the fresh catalyst to the minimum fluidization regime with the grown catalyst particles. The extent of particle growth and fluidization regime is dependent on operating conditions and time.

In the fluidized bed reactors, the scale of the effective phenomena can vary from; the mass and heat transfer within a single porous particle, the chemical reaction leading to particle growth to the interactions of particles with each other or with the fluid. Computational Fluid Dynamics â€" Discrete Element Method (CFD-DEM) is coupled with the Multi-Grain Model (MGM) to capture all these effects and enables us to study dense gas-particle flows in combination with coupled mass and heat transfer and catalytic reactions in growing particles. The obtained knowledge from this study will be used to study the TCD and to bring the industrial scale TCD in gas-fluidized beds one step closer to reality.

Pyrolysis and in Line Oxidative Steam Reforming of Hdpe

Authors

Lopez, G. - Presenter, University of the Basque Country

Suárez Cardona, M. A.

Orozco, S., University of the Basque Country UPV/EHU

Olazar, L., University of the Basque Country UPV/EHU

Cortazar Dueñas, M., University of the Basque Country

Comendador, P., University of the Basque Country UPV/EHU

Amutio, M., University of the Basque Country

Olazar, M., University of the Basque Country

The environmental concern associated with waste plastics is growing due to the persistent impact of plastics on oceans, freshwater and terrestrial environments. Within this scenario, the development of feasible processes for waste plastic management is an urgent matter to which increasing attention has been paid in recent years. Thermochemical conversion technologies for plastics are promising alternatives, as they allow producing fuel and high value-added chemicals. Recently, the combination of pyrolysis and in line steam reforming was proposed as a new strategy for the selective H_2 production.

Pyrolysis and in line oxidative steam reforming for H_2 production from high density polyethylene (HDPE) was carried out in a conical spouted bed reactor (CSBR) and a fluidized bed reactor (FBR). Operating under oxidative conditions can reduce or even avoid the energy requirements of the endothermic reactions of the reforming stage, as well as contribute to the in situ combustion of the coke deposited on the catalyst, and therefore improve its stability.

A parametric study was carried out in a wide range of operating conditions: reforming temperature (550, 600, 650, 700, 750 °C); the steam/plastic ratio (S/P) (2, 3, 4, 5) and the equivalence ratio (ER, ratio between the oxygen introduced in the reforming and that required for complete combustion) (0, 0.1, 0.2, 0.3).

In general, the S/P ratio is the most influential parameter on H_2 production, followed by ER and temperature. Thus, H_2 yields of around 24 wt% are obtained with S/P values of 2 and 3; H_2 yields of around 28 wt% with ER between 0 and 0.1 and 24 wt% with ER between 0.2 and 0.3; and H_2 yields of 18-23 wt% when the reforming temperature is between 550-650 ºC and around 25 wt% when it is between 700-750 ºC.

Roles of Fluidization Technology in Thermal Engineering for a Sustainable Future

Authors

Tsutsumi, A. - Presenter, The University of Tokyo

Over the last decade, the self-heat recuperation technology based on the exergy recuperation for innovative energy saving in which process heat is recuperated and recycled without adding heat has been developed and applied to various energy-intensive thermal and reaction processes such as drying, CO_2 chemical absorption, sea water desalination, chemical reaction systems, resulting in the reduction of energy consumption by a factor of 1/3-1/12 compared to conventional heat recovery. Self-Heat Recuperation Technology has been proved to be an effective method to cut the huge energy consumption in the chemical industry because of heat circulation. The minimum exergy required for heat circulation can be quantified by calculating the waste heat via exergy destruction due to heat transfer.

Fluidized beds have an excellent ability to rapid transfer heat and maintain a uniform temperature. Thus, the self-heat recuperative heat circulation system in the energy intensive process by using fluidized bed has a great potential for the energy saving. In this paper, the thermodynamic mechanism for exergy recuperative drying and evaporation processes by using fluidized bed is elucidated in terms of exergy analysis with thermodynamic identity and energy conversion equation in the vector notation for energy with exergy and anergy and the temperature â€" entropy (*T-S*) diagram which can express the inequivalence between work and heat and the irreversibility of the process due to the entropy generation. In addition, the exergy recuperative H_2 production system by using a internal circulating fluidized bed for the biomass gasification will be described.

Gas-Solid Flow Study of Multistage Fluidized Bed Red Mud Reduction Reactor with Downcomer

Authors

Kong, H. - Presenter

Yang, H., Key Laboratory for Thermal Science and Power Engineering of Ministry of Education, Department of Energy and Power Engineering, Tsinghua University

Zhou, T., Key Laboratory for Thermal Science and Power Engineering of Ministry of Education, Department of Energy and Power Engineering, Tsinghua University

Huang, Z., Tsinghua

Zhang, M., Key Laboratory for Thermal Science and Power Engineering of Ministry of Education, Department of Energy and Power Engineering, Tsinghua University

Multistage fluidized bed reactors with downcomers are widely conducted for red mud reduction due to their excellent residence time distribution and multiple dense phase zones, which result in favorable reaction characteristics. Furthermore, in the fluidized bed reactors, gas-solid flow plays an essential role in chemical reactions. However, the complex structure of multistage fluidized bed reactor make the study the gas-solid flow challenging, especially when the particles have an extremely fine size.

In this paper, a cold-test rig for a multistage fluidized bed cold test rig was constructed. Using experimental methods, appropriate size and structure of downcomers were determined to ensure the rapid transfer of solid between the upper and lower bed. Additionally, the distribution of gas residence time in multistage fluidized bed was obtained by using CO_2 tracer and function fitting noise reduction. The average gas residence time is 13.72 s and 50% longer than the single stage bed which is theoretically 9.09 s.

We Want You to Talk during Class: A New Graduate Course on the Oral Communication of Science

Authors

Hrenya, C. - Presenter, University of Colorado

As evidenced during the covid pandemic, scientific results and discoveries may be ignored or even misrepresented in public discourse (by media, political leaders, etc.), leading to less-than-ideal policy decisions and personal choices. To improve this situation, the training of scientists and engineers to communicate effectively to the general public, and not just their peers, is critical. In this work, a new graduate course on the oral communication of science will be described in which an emphasis was placed on making engaging presentations to a more general audience. This 1-credit course was offered remotely to graduate students during the covid lockdown. General course structure, topics covered, and lessons learned will be presented.

Bayesian Optimization of Deep Learning Algorithms in Sorption Processes

Authors

Skrobek, D. - Presenter

Krzywanski, J.

Zylka, A., Jan Dlugosz University of Czestochowa

Grabowska, K., Jan Dlugosz University of Czestochowa

Sosnowski, M., Jan Dlugosz University of Czestochowa

Kulakowska, A., Jan Dlugosz University of Czestochowa

Nowak, W., AGH University of Science and Technology

Ashraf, W. M., University College London

Czakiert, T., Czestochowa University of Technology

Gao, Y., East China University of Science and Technology

The use of machine learning algorithms often involves careful tuning of the learning parameters and hyperparameters of the model. Unfortunately, this tuning is often very difficult and requires a lot of experience. Therefore, there is a high demand for automated approaches that can optimize the performance of any learning algorithm for a given problem.

Long Short-Term Memory (LSTM), Bidirectional Long Short-Term Memory (BiLSTM) and Gated Recurrent Unit (GRU) are among the most popular deep learning models currently in use. The performance of LSTM, BiLSTM and GRU is highly dependent on the selection of several hyperparameters that must be carefully selected to obtain good results. In this work, we consider this problem through the framework of Bayesian optimization.

The article presents deep learning methods used to predict the mass of an adsorption bed in a fluidized bed. The purpose of using this type of bed is to improve the efficiency of adsorption cooling systems by increasing heat and mass transfer through the use of fluidization. The article presents numerical studies on mass prediction using the above-mentioned algorithms for silica gel as a sorbent. In selected neural networks, the following hyperparameters were optimized: number of hidden layers and number of neurons in each layer.

The described research is part of the work planned in the project No. 2018/29/B/ST8/00442, "Research on methods of intensifying the sorption process in modified adsorbent deposits", financed by the National Science Center, Poland.

Mathematical Modelling of a Bubbling Fluidized Bed Gasifier for High Ash Indian Coal

Authors

Singh, K. - Presenter, Indian Institute of Technology Delhi

Roy, S., Indian Institute of Technology Delhi

Fluidized bed gasifiers are used in a large number of process industries such as petrochemical, refinery, agrochemical, fine chemicals, and fertilizers. Formation of synthesis gas ($CO + H_2$) from coal is a viable method, as it produces the â€œbuilding blockâ€ molecules for the further down streaming process which can be used to produce both fuel and chemicals such as Methanol and Dimethyl Ether (DME). Various technical challenges faced while using high ash Indian coal includes erosion in the process equipmentâ€™s, poor emissivity and flame temperature, and low radiative heat transfer. In order to make the process commercially viable we need to develop an efficient system to handle Indian coal and find a scalable way for the generation of synthesis gas. Thus, in this study, 1D non-isothermal steady state mathematical model was developed for gasification of â€œIndian coalâ€ in a bubbling fluidized bed coal gasifier by incorporating steady state mass conservation equations for â€œtotal flowâ€ of gaseous species along the bed height. This model incorporates the effect of hydrodynamic behaviour of the bed and complex reactions such as coal devolatilization, homogeneous, and heterogeneous reactions. Steady state energy balance equations for solid fuel, gas phase and inert bed material were written to predict the average temperature variation along the height of gasifier. Continuity equation for the variation of superficial gas velocity and other hydrodynamic parameters such as bubble diameter and bubble rise velocity were also written. Pre-exponential factor for both homogeneous and heterogeneous reactions were fitted by using genetic algorithm to match the exit synthesis gas composition obtained from actual Indian coal gasifier. The simulation results thus obtained were in good agreement with the results obtained from actual Indian coal gasifier. The proposed model can be used to scale-up, operational control, or improving the design of any bubbling fluidized bed gasifier.

Modulation of Turbulence By Dispersed Charged Particles in Pipe Flow

Authors

Yao, J. - Presenter, China University of Petroleum-Beijing

Zhao, Y., China University of Petroleum

Liu, M., China University of Petroleum (Beijing)

Turbulent gasâ€“solid two-phase flow with electrostatic effects is studied. The turbulent pipe flow is treated using large eddy simulation (LES), while the particles are tracked using a Lagrangian approach. Simulations are carried out with one-way coupling, two-way coupling, and two-way coupling taking account of electrostatics, and the results are compared. The bulk Reynolds number Re_b = 44 000 and the Stokes number St = 3.9 (d_p = 5 Î¼m). The results show that the maximum electrostatic field strength in the saturated state is found near, but not at, the wall. The electrostatic effect increases the particle concentration in the viscous sublayer (0 â‰¤ y^+ â‰¤ 5) and the particle dispersion in the buffer layer (5 â‰¤ y^+ â‰¤ 30). Owing to the electrostatic effect, the feedback effect of particles on the fluid is increased, which leads to increases in the average fluid velocity in the buffer layer and in the velocity fluctuations. In addition, the electrostatic effect is found to increase the turbulent kinetic energy near the wall, while this trend decreasing with distance away from the wall. The areas of high- and low-speed streaks near the wall are increased by the electrostatic effect. Therefore, it can be concluded that electrostatics changes not only the particle behavior, but also the flow field.

Plastics-Bed Interaction in Industrial Fluidized Bed Reactors during Thermochemical Conversion Processes

Authors

Iannello, S. - Presenter

Materazzi, M., University College London

Developing economic and environmentally acceptable solutions for managing the ever-increasing volumes of plastic waste that are generated worldwide is one of the greatest challenges of our time. Although mechanical recycling clearly has a critical role in reducing the amount of waste to be disposed, there is further opportunity and environmental benefit in what is called â€œchemical recyclingâ€. Advanced thermochemical technologies, like pyrolysis and gasification, in fluidized bed reactors have an important role to play in converting these feedstocks into valuable intermediates (e.g., methanol, DME and hydrogen) for manufacturing industries. Nevertheless, there are still unsolved challenges when operating with mixed plastic waste feedstocks. These materials are highly heterogeneous and tend to sink soon after injection to the bed, due to the formation of lumps and agglomerates. This can cause segregation, de-fluidization of the bed and, eventually, early shutdown of the operation.

The aim of this work is to link the thermal decomposition behaviour of plastic to its interaction with bed materials at a range of operating conditions relevant to industrial operations. In particular, non-invasive X-ray imaging and infrared gas measurements have been used to investigate plastic-bed interaction and rate of decomposition of a single polypropylene particle in a lab scale facility. A separate experimental campaign is conducted to record the temperature profile inside the particles during the transformation process by inserting a thermocouple into half-drilled plastic samples. This information is essential to understand the thermal history and hydrodynamic behaviour of plastic materials during thermal degradation in fluidized beds, whose details, unlike traditional thermogravimetric data, are still uncertain. The present study highlights the importance of gathering new observations to better understand the fundamental phenomena which govern the interaction between solid feedstocks and fluidized bed, hence promoting the exploitation of such technology in an industrial framework.

Effect of Reaction Temperature and Regeneration Temperature of DME Carbonylation to MA Reaction on Regeneration of Spray-Drying Ferrierite Catalyst

Authors

Sung, W. - Presenter

Lee, D. H., Seintec Corporation

Jung, H. S., Sungkyunkwan University

Bae, J. W., Sungkyunkwan University (SKKU)

Kim, J., Incheon National University

Kim, J. Y., The University of British Columbia

The effect of reaction temperature and regeneration temperature of DME carbonylation to MA reaction on regeneration of spray-drying ferrierite catalysts was investigated. A zeolite-based spray drying ferrierite (S-FER) belongs to the Geldart A classification (d_p =51.2 Î¼m, Ï?$_s$ =1029 kg/m^3) and 0.01 kg of S-FER was loaded in the reactor (I.D. = 0.0243 m, h= 0.5 m). The reaction was conducted by changing the reaction temperature (240-270â„ƒ) while fixing the pressure = 50 bar, the reaction gas volume fraction $CO/DME/N_2$ = 0.2/0.01/0.79, space velocity = 1250 L/kg_{cat}/h, and experiment time = 20 h. For the regeneration, air was used, and regeneration temperature was changed from 300- 600â„ƒ, while fixing space velocity = 1000 L/kg_{cat}/h, pressure = 50 bar, regeneration time = 6 h, temperature ramping rate= 1Â°C/min. Although reaction temperature was changed from 240â„ƒ to 270â„ƒ, recovery of catalysts activity was not changed significantly. On the other hand, the regeneration temperature had a great effect on the recovery of catalysts recovery. When the regeneration temperature was 600â„ƒ, the recovery of catalysts activity was 43%. However, the recovery of catalysts activity close to 99%, when the regeneration temperature was 500â„ƒ.

Proof of Concept Calcination Kinetic in TGA and Fluidised Bed Reactor

Authors

Boege, K. - Presenter

Karl, J., Friedrich-Alexander-University Erlangen-Nuremberg

Calcination of limestone releases large amounts of CO_2, 60 % by the limestone itself. In order to capture CO_2, the indirect fluidised-bed calcination is a suitable option where CO_2 is the fluidisation medium. However, there are also challenges as the fluidisation with CO_2 encourages sintering and requires temperatures above 900 °C in order to achieve complete decomposition. An alternative is steam, which lowers the CO_2 partial pressure and has a catalytic effect on the calcination reaction.

Steam contents up to 60 % are well studied in the lap scale by thermogravimetric analysis (TGA). The study is intended to close the gap between TGA experiments and the steam calcination in an up-scaled fluidised bed under consideration of a 100 % steam influence. Experiments are carried out to determine the kinetics of calcination (0-100 % H_2O/N_2 in CO_2, T = 700- 950 °C) by TGA and the fluidised bed reactor. The fluidised bed is interspersed with electrically heated tubes, which provide the required heat (20 kW) for the 18 kg reactor content.

For two particle size distributions (100-315/300-700 µm) it can be shown that the start of the decomposition temperature sinks from 900 °C at 100 % CO_2 to 740 °C at 100 % H_2O. As the CO_2 partial pressure increases over the height of the fluidised bed due to the CO_2-release there is not a 100 % steam atmosphere in the reactor even with 100 % steam fluidisation in the inlet. Contrary to expectation the residence times (40-60 min, >95 %, at 900 °C) is the same for both scales.

Finally, the samples are analysed for reactivity using the T-60 test, which is well-known in the lime industry, in order to create a link to the product-quality and the reactivity of lime.

Large-scale thermochemical energy storage using high-temperature solid cycles â€“ impact of electricity system composition

Authors

Toktarova, A. - Presenter

Martinez Castilla, G., Chalmers University of Technology

Guío-Pérez, D. C., Universidad Nacional de Colombia

Pallarès, D., Chalmers University of Technology

Johnsson, F., Chalmers University of Technology

The replacement of fossil fuels with electricity from renewables is a main option in climate change mitigation. Thus, future electricity systems are expected to consist of a large share of variable renewable electricity (VRE) and these will require flexibility measures to be applied in order to increase the value of VRE, i.e., to reduce VRE curtailment.

Energy storage can offer flexibility since it can disconnect electricity generation from the demand, and thermochemical energy storage (TCES) using high-temperature solid cycles appears as a technology with the potential to store energy at different time scales (from hours to seasons). The use of fluidized bed (FB) reactors allows for implementation of TCES at a large scale.

This work investigates how large-scale TCES influences the cost-optimal investment and operation of electricity and district heating (DH) systems in the electricity price area of southern Sweden. TCES offers the possibility to absorb non-dispatchable electricity and store it as chemical energy (in the form of reduced solid material), which can subsequently be discharged as dispatchable heat. Based on previous studies at process-level, this work assumes iron oxide as the redox material and green hydrogen as the reducing agent for the chemical reduction.

A previously developed energy system cost optimization model is further developed to also account for heat production flexibility including TCES with high-temperature solid cycles. The model minimizes the investment and operating costs of electricity generation and heat production units while meeting the demands for electricity and heat without adding carbon dioxide to the atmosphere.

The results show that the extent to which the cost of electricity for heat production can be reduced depends on the operational flexibility of the heat production units and the extents of hydrogen and solids storage. To achieve a low electricity cost for heat production, investments in overcapacity are required.

Comparison of Drag Models for Modeling Fluidized Beds with the Coarse Grain DEM-CFD Approach

Authors

Grabowski, J. - Presenter, Technical University Berlin

Jurtz, N., Technical University Berlin

Kraume, M., Technical University Berlin

CFD coupled with DEM is widely used for accurate simulations of fluidized beds. However, the CFD-DEM approach is computationally very demanding. As for today, it is restricted to lab-scale reactors with only a limited number of particles. Coarse grain approaches with coupled CFD-DEM have become popular over the last years to address the aforementioned issue. Particles of the original system are summarized into so-called grains. Thus, the number of particles can be reduced drastically, allowing the calculation of industrial-scale reactors in a realistic time frame. An important parameter is the coarse grain factor, which is defined as the ratio of grain diameter to particle diameter:

$l=d_g/d_p$

The goal of the coarse-grain approach is to mimic the original system precisely. However, with growing grain size, accurate representation of the interaction forces becomes challenging. Coarse simulations can fail to resolute the sub-grid phenomena, such as bubble and cluster formation, which can lead to overprediction of drag force.

As a solution to this, sub-grid drag models can be applied. Sub-grid drag models are developed by regressing finely resolved simulations, minimizing the energy for transporting particles in the flow, or correlating experimental results. The models, however, tend to be ad-hoc, and a thorough analysis of the limitations of the range of process and system parameters is often missing. In this work, common sub-grid drag models are compared against finely resolved DEM-CFD simulations. In the first step, the limitations of using the coarse grain approach with a homogeneous drag law are investigated. In the second second step, inaccurate simulations are repeated with selected sub-grid models and their applicability is evaluated for fluidized bed configurations with different Geldart-type particles, gas inlet velocity, and reactor width.

Gasification Characteristics of Disposable COVID-19 Masks Using a Bubbling Fluidized Bed: Effect of Gasifying Agents

Authors

Nam, J. Y. - Presenter

Seo, M. W., University of Seoul

Tokmurzin, D., Korea Institute of Energy Research

Lee, D. H., Seintec Corporation

Ra, H. W., Korea Institute of Energy Research (KIER)

COVID-19 and mutated viruses have increased global demand for masks, resulting in 490,000 tons of trash masks every month. Waste masks include a virus infection risk and are difficult to recycle due to their multi-component polymer and iron composition. Conventional treatment methods, such as landfilling and incineration, can emit environmental pollutants and cause respiratory diseases. Gasification is a reasonable method for treating multi-component polymers and producing syngas for a variety of applications. The gasification characteristics of disposable COVID-19 masks using a 1kg/h laboratory-scale bubbling fluidized bed gasifier were determined in this study. The experimental operating temperature was set to 1073 K, the fuel feeding rate was set to 6.95 x 10^{-5} kg/s, and the minimum fluidization ratio (U_{mf}/U_g) was set to be 2.5. The characteristics of the waste mask thermochemical conversion process and the influence of activated carbon were investigated depending on the presence or absence of a gasifying agent and the type of gasifying agent (air, steam, carbon dioxide, oxygen/steam). The hydrogen production yield shows maximum under pure oxygen/steam gasification conditions in the presence of activated carbon, and the H_2/CO molar ratio was highest in steam gasification condition. The maximum product gas production was obtained by air gasification, while carbon dioxide gasification was shown to be effective in reducing tar. In the most optimal hydrogen production condition, the gas composition is H_2 33.4 vol%, CO 13.1 vol%, CH_4 16.0 vol%, CO_2 19.1 vol%, and hydrocarbon (C_2-C_3) 18.4 vol%. Activated carbon had the highest tar reduction efficiency of 91%, while CO_2 had the highest tar reduction efficiency of 49% among gasifying agents. These results provided the design parameter for waste mask gasification process, and the product gas can be used as feedstock for chemicals and fuels in the near future.

Understanding Features in Fluidisation Curves for the Evaluation of Powder Cohesivity

Authors

Thomas, A. - Presenter, Freeman Technology

Vriend, N., University of Colorado

Routh, A., University of Cambridge

Features of typical pressure drop and voidage curves formed during the fluidisation and de-fluidisation of powders are compared for a range of powders with different flow behaviours. A hysteresis in the pressure drop across the powder bed, an overshoot in pressure drop around the minimum fluidisation air speed, and the fraction of the bed weight suspended by drag are observed to correlate with traditional metrics of cohesivity. Fluidisation tests were carried out in an FT4 Powder Rheometer, where a standard Aeration Test was modified to track the pressure drop and the bed expansion during fluidisation at appropriate air speed intervals. Based on the hysteresis in pressure drop and voidage curves, we propose a novel method for experimentally determining the Bond number, i.e. the ratio of average interparticle force magnitude to particle weight. The difference between the fluidisation and de-fluidisation curves is related to the force required to break interparticle forces in the powder bed between two packing states. By computing the change in coordination number (mean number of contact points per particle) estimated from the bed void fraction, the number of interparticle forces broken during aeration is related to the average particle-particle force. The results are compared to theoretical expectations and also to alternative rheological measurements commonly used in industry, including the Hausner ratio, the angle of repose shear cell cohesion and flow function values, and FT4 Energy measurements. Furthermore, the proposed method is shown to be independent of bed dimensions and initial packing state.

Radar-Based Measurements of the Solids Flow in a Circulating Fluidized Bed - Back-Mixing and Clustering

Authors

Wu, W. - Presenter

Duan, L., Southeast University

Pallarès, D., Chalmers University of Technology

Johnsson, F., Chalmers University of Technology

Guío-Pérez, D. C., Universidad Nacional de Colombia

Bonmann, M., Chalmers University of Technology

This work provides an experimental investigation of the solids back-mixing in the freeboard of circulating fluidized beds. The use of radar-based measurements (a novel technique for fluidized beds) provides detailed data on the velocity distribution of the solids phase, which allows quantification of the two main solids back-mixing mechanisms: clustering mainly in the core region of the riser and downward flow of solids at the riser walls forming wall layers. The experiments are carried out in a circulating fluidized bed riser with a height of 3.1 m and 0.45 m^2 in cross section operated with Geldart B solids at room temperature and involve both radar diagnostics (for the solids velocity distributions) and pressure sampling (for the vertical profile of solids concentration). With this data, mass balance closure of the solids flow can be applied from which both the mass transfer coefficient for the core-wall layer exchange and the volumetric rate of cluster formation/destruction can be derived. Having this, the effect of flow parameters (gas superficial velocity and solids concentration) on the transfer of solids from the core to the walls and the clustering rate are investigated and discussed.

Keywords: Circulating fluidized bed; Radar technique; solids mixing; Discrete phase flow; Cluster phase flow

Accuracy Improvement of Numerical Simulation of a Hydraulics with Wood Pellet Combustion Operating in a Full-Loop Circulating Fluidized Bed Combustor

Authors

Park, S. S. - Presenter, Korea Institute of Energy Research

Moon, J., Korea Institute of Energy Research

Three-dimensional computational fluid dynamics (CFD) modeling of the hydrodynamic characteristics for the various sand particle sizes (10 mm, 250 mm, 482 mm, 750 mm, 1000 mm, and the actual PSD (particle size distribution)) in a 0.1 MW_{th} pilot scale full-loop circulating fluidized bed combustor (CFBC) was numerically investigated with a commercial CFD code, ANSYS FLUENT. Those were clearly validated with actual CFBC operating data. As the result, the sand particle size of the actual PSD were in good agreement. In order to accurately simulate wood pellet combustion, kinetic parameters for activation energy (E_a) and pre-exponential factor (A) for wood pellet decomposition and char combustion were obtained from lab-scale TGA experiments with wood pellet sample actually used for CFBC operation. The CFD modeling on wood pellet combustion were conducted with the selected sand particle size of actual PSD with improved char combustion model using the user-defined function (UDF) reflected kinetic parameters. The CFD modeling results were carefully compared with actual CFBC operating results. Therefore, our CFD simulation results for the concentrations of gas species (N_2, CO_2, O_2, SO_2, CO, NO, and N_2O) at the gas exit in the cyclone were also in good agreement with the actual CFBC operating data. The accuracy improvements were increased from 1.9 times to 2.5 times due to our improved char combustion model using the UDF.

Floating and Sinking Behaviour of a Heat Transfer Sphere Submerged in a Gas-Solid Fluidised Bed

Authors

Garcia-Trinanes, P. - Presenter, University of Greenwich

One of the most important features of a fluidised bed is its good heat-transfer characteristics. The role of various parameters and the mechanism of heat-transfer in fluidised beds have been a subject of extensive investigations in the recent past [1,2]. Oshitani et al. [3] reported that under certain conditions, a sphere can sink or float depending on the relative density. This phenomenon has been attributed to a local fluidization around the sinking sphere. Assuming this is correct, the heat transfer on such a sphere is therefore affected too. In this study a metallic sphere is used to monitor the heat transfer coefficient in such conditions involving as it does a few inter-dependent variables. The position of the sphere has been also monitored giving further information about the hydrodynamics of the system. This study has industrial relevance as it affects dry separators, catalytic cracking or thermochemical energy conversion processes.

1. García-Triñanes, J.P.K. Seville, R. Ansart, H. Benoit, T.W. Leadbeater, D.J. Parker, Particle motion and heat transfer in an upward-flowing dense particle suspension: Application in solar receivers, Chemical Engineering Science, Volume 177, 2018, 313-322, htps://doi.org/10.1016/j.ces.2017.11.041.

2. Mayumi Tsukada and Masayuki Horio, Maximum heat-transfer coefficient for an immersed body in a bubbling fluidized bed, Industrial & Engineering Chemistry Research 1992 31 (4), 1147-1156 htps://doi.org/10.1021/ie00004a025.

3. Jun Oshitani, Toshiki Sasaki, Takuya Tsuji, Shusaku Harada, Hirokazu Kajiwara, Kei Matsuoka, Unstable sinking of spheres at higher air velocity in a gas-solid fluidized bed, Advanced Powder Technology, Volume 32, Issue 4, 2021, 1300-1304,https://doi.org/10.1016/j.apt.2021.02.021.

Direct Measurement of Apparent Viscosity of a Fluidized Bed Under Non-Ambient Conditions

Authors

Schütz, D. - Presenter, Anton Paar Gmbh

Pondicherry, K., Anton Paar GmBh.

The rheological behavior of fluidized beds has been an object of study ever since their first technological use in the 1940´s.

However, the study of their behaviour under non ambient conditions is severely limited.

While strides are being made in trhe last years in the characterization of hot and cold granular media in the consolidated state this work focuses on the direct measurement of apparent viscosity and other significant parameters of fluidized beds under direct application of non-ambient conditions made possible with a modification of our powder flow cell.

The focus will be on polymeric powders due to their widespread use as well as the significant change they experience with the application of both high and low temperatures.

This will be contrasted with the behaviour of both the bulk material measured by Dynamic mechanical analysis as well as their consolidated granular behaviour with a heated shear cell.

Hydrodynamic Study of a Bench Scale Fountain Confined Conical Spouted Bed Reactor for Biomass Torrefaction

Authors

Sukunza, X. - Presenter

Bolanos, M., University of the Basque Country (UPV/EHU)

Tellabide, M., University of the Basque Country

Estiati, I., University of the Basque Country

Aguado, R. Sr., University of the Basque Country

Olazar, M., University of the Basque Country

Torrefaction of biomass is a mild pyrolysis process conducted at 200 â€“ 300 ÂºC in an inert atmosphere, which allows attaining high energy density solid feedstock. Accordingly, torrefaction of biomass is commonly carried out in a great variety of reactors, such as rotary drums, and fixed and fuidised beds. Furthermore, spouted beds are a promising alternative, as they allow operating with large particle size materials, such as those of biomass. However, continuous torrefaction of biomass in spouted beds is commonly conducted using beds of inert materials. Thus, the biomass is fed from the top of the reactor and the torrefied biomass is entrained through the top outlet pipe. Nevertheless, this configuration leads to coarse particle accumulation in the bed. Accordingly, new alternatives are required in continuous operation, preferably those involving beds made up of only biomass and high solid circulation. Thus, the presence of a fountain confiner on top of the bed allows increasing the gas-solid contact, with the torrefied biomass being removed through a solid outlet pipe located at the same level as the bed upper surface.

Therefore, this study deals with the performance of a fountain confined conical spouted bed, in which hydrodynamics was evaluated using different biomasses (rice husk, almond shell and olive stone). Moreover, the effect of the draft tube and the fountain confiner was analysed and the configurations leading to stable operation were identified. Furthermore, the minimum spouting velocity and operating pressure drop were measured. The hydrodynamic behaviour of beds made up of almond shells has clear similarities with those of olive stones, which is explained by the common features of these solids. Overall, the results allow delimiting the operating conditions for biomass torrefaction in fountain confined conical spouted beds.

Understanding Particle Flows in Sub-Fluidized Horizontal Stirred Bed Reactors By Radioactive Particle Tracking

Authors

van der Sande, P. C. - Presenter, Delft University of Technology

Van Ommen, J. R., Delft University of Technology

M.H. Meesters, G., TU Delft

Horizontal stirred bed reactors are commercially employed in polypropylene manufacturing processes. The reactors generally contain a bed that is mildly stirred by a series of paddles attached to a center shaft. Via nozzles located at the bottom of the reactor gaseous propylene enters the reactor at controlled flowrate to prevent bed fluidization, bringing the bed in sub-fluidized state. It is crucial to have a well-mixed system with a narrow particle cycle time distribution throughout the reactor to avoid the formation of hotspots that could lead to expensive reactor shutdowns due to lump formation. It is therefore of great importance to understand the particle flows in horizontal stirred bed reactors.

In this research, we developed a lab-scale horizontal stirred bed reactor that enables particle flow studies in non-reactive environments. Since optical techniques are inadequate to study the dense flows, we study the reactor hydrodynamics with X-ray imaging and radioactive particle tracking (RPT). Gas-solid distribution maps acquired with X-ray imaging showed that at higher inlet velocities bed spouting occurs, leading to substantial gas by-pass and poor gas-solid interaction. Now we use RPT to characterize particle trajectories in the reactor. The RPT setup consists of three scintillation detectors that form a field of view around the reactor. Within this field of view the location of the tracer particle is reconstructed with sub-millimetre and sub-second accuracy. We studied the effect of the reactor filling, gas inlet flowrate and agitator rotation speed on the particle cycle time distribution and axial dispersion for industrial-grade polypropylene (Geldart B type). We found that at higher reactor fillings and higher rotation speeds the particle has a shorter cycle time, which would lead to improved heat removal in the process. The results from this study contribute to better understanding of the flow behaviour in horizontal stirred bed reactors.

FLUIDIZATION IN NATURE: PYROCLASTIC FLOWS

Authors

Salatino, P. - Presenter, Università degli Studi di Napoli Federico II

Pyroclastic flows are mixtures of gas and particles generated by gravitational collapse of lava domes or fall-back of eruption columns during explosive volcanic eruptions. They represent major natural hazards because they can propagate far away from the source at speeds that may largely exceed those of most natural or anthropogenic granular flows. Hazard mitigation requires the development of reliable models based on mechanistic understanding of their complex dynamics to predict flow path, velocity, momentum, and run-out distance.

It has long been hypothesized that fluidization of pyroclasts - by either endogenous gases or entrainment of ambient air - may represent one key to the emplacement of dense pyroclastic flows, and a key factor for their extraordinary mobility. Although early speculations on the role of fluidization have received substantive support, the debate is still open as to whether the apparently inviscid nature â€" hence large mobility â€" of pyroclastic flows arises mostly from fluidization or from inherent rheological patterns of fast dry granular flows. Broad uncertainties characterize fundamental aspects like the prevailing nature and source of the fluidizing gas, the extent of particle-phase stresses in rapid aerated/fluidized dense granular flows, the perturbation to classical fluidization regimes due by turbulence and shear, the influence of solids polydispersity, the dynamics of the frontal zone of the flow and its relevance to gas entrainment. The current lack of fundamental understanding, despite the extensive published literature on the subject, is largely due to limitations of fundamental research, mostly based on scaled-down physical models of granular currents.

The lecture will survey and critically discuss the literature relevant to the subject, with the aim of establishing an up-to-date mechanistic framework of pyroclastic flows embodying consideration of pyroclast fluidization.

Stochastic Modeling of Dilute and Moderately Dense Particle Flows for Large Scale Systems

Authors

Morris, A. - Presenter, Purdue University

Hong, A., Purdue University

The direct simulation Monte Carlo method is used to simulate particle flows by stochastically solving the governing revised Enskog equation. Unlike deterministic Lagrangian modeling approaches, collisions between particles are simulated stochastically. Nearby particles are randomly selected to collide based on a kinetic theory-based collision probability. After the collision, particles can scatter and dissipate energy via inelastic losses. By performing the collisions stochastically, the DSMC approach is shown to be more than an order of magnitude faster than deterministic modeling approaches. Although the traditional DSMC method is able to accurately capture the solid stresses in dilute granular flows, traditional DSMC does not accurately capture the collisional and virial stresses when the solid concentration becomes moderately dense. As a result, traditional DSMC is susceptible to inelastic collapse, and this phenomenon can lead to unphysical over-packing. We find that the virial pressure is not recovered in traditional DSMC because the convection of particles does not consider volume exclusion effects. To include volume exclusion effects in the convective phase, we implement a random walk method. It is found that by making a relatively simple correction to how particles convect, we are now able to recover the correct collisional pressures for moderately dense flows and prevent unphysical over-packing of particles. The new method is compared to coarse grained DEM and DEM simulations for a bubbling bed.

Future Perspectives of Effective Utilization of Biomass -Renewable Carbon Resource

Authors

Fushimi, C. - Presenter, Tokyo University of Agriculture and Technology

Many biomass power plants have been build and operated because biomass power generation systems are considered to be renewable. However, biomass power generation has many drawbacks, resulting in much lower thermal efficiency and costly than wind and photovoltaic power generation systems. Unfortunately, simple combustion (including the combustion of biomass) has critical problems such as i) emission of CO_2 and other pollutants, ii) great reduction of economic value of raw materials, and iii) large exergy loss. In contrast, combustion has following strong points; I) high power, II) high temperature heat generation, and III) utilization for hygiene/sanitation purposes or aftertreatments of organic wastes. The integration of energy storage system with the biomass power plant for non-steady state operation is a promising method to balance the intermittent electricity supply by variable renewable energy and electricity demand. Pyrolysis and gasification of biomass (including organic wastes) with fixed/fluidized beds is an important technology for carbon recycle and biocarbon production.

Transformation of Plastic into the Liquid Fuel

Authors

Upadhyay, K. - Presenter, PKG Group of Institution

Abstract

Globally, over 380 million metric tons plastics are produced in a year and the bins are also overflowed because of used plastics. Presently, there are many technologies which recycle the plastic waste to support in controlling pollutions. They give the various drawbacks and after effects to ecological systems, like non biodegradability, enter of toxic compounds in food chain, etc. In this era, we may use biodegradable plastics or we can take conclusive steps for decomposing the non biodegradable plastics. Recycling methods for non biodegradable plastic wastes must be according to a conventional strategy. But recycling technique is not having an excellent result, as cleaning or decomposition of plastic wastes along many difficulties. Removal of non-biodegradable plastics by landfill methods, blast furnace, gasification and incineration, are not the best solution. That produces toxic gases along high cost of production / processing.

From chemical recycling method, like pyrolysis, is the one of the best or less time consuming method of decomposing non biodegradable plastics wastes. Pyrolysis of non biodegradable plastics is a best technique for producing alternate fuel, to protect our environment and also to save foreign exchange. This process involves the catalytic degradation of plastic waste to generate a liquid fuel, a range of hydrocarbons, petro-chemicals, etc. A catalytic cracking process, in which the plastic is cracked at high temperature and the produced gas is condensed into a liquid. That liquid is to be upgraded in to a standardized fuel by additives blending or other modifications. The type of plastic is also effect on the conversion rate of plastic into fuel and the results of this process are found to be better than other methods for decomposing of plastics because pyrolysis has better yield of conversion in to useful products.

Application of Fluidized Bed Reactors for Carbon Capture

Authors

Duan, L. - Presenter, Southeast University

Tang, H., Georgia Institute of Technology

Sun, Z., Natural Resources Canada

Wang, Y., University of Utah

Duan, Y., Southeast University

Li, L., Southeast University

Zhu, C., Southeast University

With the aim to mitigate the global warming, carbon capture, utilization and storage (CCUS) technology is expected to play a pivotal role in the decarbonization of industrial production and the achievement of a carbon-neutral future, thus attracting broad research interests. In particular, the promise of fluidized bed reactors for CO_2 capture from large industrial sources has been demonstrated by a suite of CCUS technologies. This review provides a perspective on CO_2 capture technologies that employ fluidized bed rectors, involving oxy-fuel combustion, chemical looping combustion, calcium looping, and low-temperature sorption. Based on an overview from the lab-based fluidized bed reactors to pilot-scale demonstration facilities, the state-of-the-art progress summarized and remaining challenges are discussed. Further research orientation of fluidized bed reactors for CO_2 capture is also outlooked in the end. We anticipate endeavors made in this review can impel the commercialization of fluidized bed-based reactors for CO_2 capture from large industrial source.

KEYNOTE: Fluidization in the Time of Sustainability Transition and Consilience

Authors

Horio, M. - Presenter, Tokyo University of Agriculture and Technology

The high time of fluidization engineering was ignited by the advent of petroleum utilization for automobile and aviation fuels then further strengthened by many industrial applications including chemical and plastics production, incineration, pyrolysis, combustion, gasification, drying, agglomeration, coating, etc. However, the linear foresight based on such a success story may become ineffective for the research and technology development in the time of climate crisis and sustainability transition. Here, it has to be recognized that the climate crisis is nothing but a clear indication that the Earth entered into the Anthropocene a new geological time. Furthermore, the time of internet and AI means that we are in the new historical time of consilience although the human and institutional cultures seem to be still in the dream of previous modernization periods.

Fluidization science and engineering have overcome sectional divisions existed in the past due to different powder characteristics and applications. Now we conceptually have the full spectrum of powder behavior from fixed/moving beds to pneumatic transport. Nevertheless, the latest knowledge that our fluidization society has developed are not well shared by other disciplines. Also, the activity of fluidization engineering has still much in the field of fuel and energy applications.

In this presentation, I am proposing a strategic approach to popularize fluidization to contribute collaborations with other disciplines, technology genres and educational circles in the time of sustainability transition and consilience, by visiting issues of definition, classification, standardization, and new challenges.

Multiphase Reactors in Solar-Assisted Fuel and Chemical Production

Authors

Li, L. - Presenter, The Australian National University

Wang, C. H., National University of Singapore

The utilization of multiphase reactors has been prevalent in various industrial applications due to their capacity to enhance heat and mass transfer, achieve high reaction conversion rates, and exhibit a high degree of reaction flexibility. Considering the growing need for sustainable solutions in the global industrial sector, there is an increasing interest in the utilization of solar energy to drive multiphase reactors. In this regard, this study presents numerical investigations of two types of multiphase reactors that are suited for solar-assisted fuel and chemical production. These include (*i*) a gas-liquid multiphase photobioreactor for microalgae cultivation operating at room temperature, and (*ii*) a gas-solid fluidized-bed reactor for solar-driven biomass gasification at approximately 900°C. Transient multiphase computational fluid dynamics (CFD) models were developed and validated against experimental results to study the gas-liquid bubbly flow in the photobioreactors and the gas-solid bubbling fluidization flow in the biomass reactor, respectively. The CFD model was further coupled with heat transfer, solar radiation transport, growth kinetics of microalgae, and biomass gasification kinetics to provide an accurate and comprehensive prediction of the complex multiphysics processes.

Acoustic Emission Techniques Combined with Machine Learning for Particle Size Distribution in Solid-Gas Fluidized Beds

Authors

Errigo, M. - Presenter, University College London

Hossein, F.

Cheng, S., Imperial College London

Angeli, P., University College London

Lettieri, P., University College London - Torrington Place

Materazzi, M., University College London

Arcucci, R., Imperial College London

The characterization of particles in solid-gas flows is of great importance in many industrial sectors such as nuclear energy and pharmaceuticals industries. Acoustic emission (AE) techniques can provide non-intrusive, multi-point measurements in non-transparent test sections and complement other techniques e.g., optical sensors, x-ray imaging. The combination of AE techniques with machine learning algorithms (ML) is particularly beneficial for industrial settings, because it reduces the cost of signal post processing and is suitable for both small and large-scale experiments. In this paper, we develop an acoustic emission technique in combination with machine learning algorithms to characterize the particle size distribution in a solid-gas fluidized bed. The theory underpinning the generation of an acoustic emission signal in solid-gas flows is explained. An AE signal is generated in gas-solid fluidized beds in the form of an elastic wave with frequencies >100 KHz and it propagates through the gas-solid mixture. An inversion algorithm to extract the information on the particle size from the energy of the AE signal is described. Experiments were carried out in a pseudo-2D flat fluidized bed with four glass bead samples with different sizes, ranging from 100 μm to 710 μm. AE signals were recorded with sampling frequency of 5 MHz. The AE signal post processing and the data preparation for the ML process are explained. For the ML process, the AE frequency, AE energy and particle velocity data were divided in training (60%), cross validation (20%) and test (20%) sets. Two ensemble ML approaches, namely Random Forest and Gradient Boosting are applied to predict particle sizes based on the AE signal features with an accuracy greater than 99.5% and a R squared value larger than 0.999. Experimental results indicate that the AE technique is a powerful tool for the characterization of the particle size in solid-gas flows.

Optimization of Heat Transfer Coefficient in Fluidized Adsorption Bed By Gene Expression Programming Approach

Authors

Krzywanski, J. - Presenter

Nowak, W., AGH University of Science and Technology

Skrobek, D.

Zylka, A., Jan Dlugosz University of Czestochowa

Ashraf, W. M., University College London

Grabowska, K., Jan Dlugosz University of Czestochowa

Sosnowski, M., Jan Dlugosz University of Czestochowa

Kulakowska, A., Jan Dlugosz University of Czestochowa

Czakiert, T., Czestochowa University of Technology

Gao, Y., East China University of Science and Technology

Adsorption cooling and desalination techniques in adsorption chillers (AC) are one of the most promising areas in energy technologies. However, the main disadvantages of conventional packed-bed ACs are the low coefficient of performance and bulkiness. These are mainly due to the sorbent beds' high voidage, leading to low heat transfer coefficients. Despite many attempts, no practical solution has been deployed to handle this problem.

To fill this gap, we proposed a novel idea to employ fluidized adsorbent bed instead of the conventional, packed bed commonly used in traditional ACs. Furthermore, gene expression programming (GEP) was introduced as a novel artificial intelligence approach to optimize the heat transfer coefficient in the adsorption bed. The current studies, which include both calculations and model validation, are performed for an intensified heat transfer adsorption bed reactor designed for low-pressure adsorption processes. The adsorption bed is fluidized with water vapour generated in the evaporator. Silica gel as the parent adsorption material and carbon nanotubes and aluminum particles as two additives with different shares were applied in the tests. The heat transfer coefficient, measured during experiments and predicted by the developed model, is investigated and compared. The data evaluated by the model agrees well with the experimental results. Simulations showed that the performed GEP-based model correctly describes the heat transfer coefficient and can be applied to study the fluidized adsorption bed reactor and optimize the bed's operating strategy.

The described investigations are a reference point for further simulations of the intensified heat transfer adsorption bed reactor since they constitute a part of the work scheduled in project No. 2018/29/B/ST8/00442, "Research on sorption process intensification methods in modified construction of adsorbent beds," funded by the National Science Center, Poland.

An Exploratory Study of Annular Fluidized Bed Reactor with High Centrifugal Acceleration for Some Industrial Processes

Authors

Li, T. - Presenter, National Energy Technology Laboratory

Pannala, S.

West, D., SABIC

Dittrich, C., SABIC

This study explored the feasibility of gas-solids fluidized bed reactor under high centrifugal acceleration for some chemical processes. The goal was to assess an ideal gas-solids contacting device with high slip velocity so that one can achieve intensified heat and mass transfer as well as the ability to fluidize hard-to-handle cohesive powders. A series of CFD-DEM simulations was first conducted for a vortex bed reactor with multiple tangential gas injections to illustrate the basic concept for group B particles in a small-scale system operated under the moving bed flow regime. The simulation results demonstrated a desired continuous operation with near-plug axial solids flow and high gas-solids slip velocity. Simple force analyses were then conducted to identify some key challenges and limitations in the design and operation of a vortex bed reactor. The analyses suggested that it would be beneficial to decouple tangential and radial driving forces for maintaining a stable centrifugal force field for reactor scale-up. A rotating fluidized bed was then studied which imposed high centrifugal force on the dense solids bed through the rotating column wall. The two-fluid model (TFM) simulations were used to carry out a virtual scale-up exercise of a horizontal rotating fluidized bed. Major challenges in the stable operation of rotating fluidized beds at large scales were discussed and a possible method to mitigate those challenges was proposed and evaluated numerically.

Experimental Investigation of Different Limestones As Sorbents for Sorption Enhanced Gasification in a Lab-Scale Dual Interconnected Fluidized Bed System

Authors

Scala, F. - Presenter, Istituto di Ricerche sulla Combustione, Consiglio Nazionale delle Ricerche

Coppola, A., P.le Tecchio 80, 80125, Napoli, Italy

Massa, F., Università degli studi di Napoli Federico II

Montagnaro, F., Università degli Studi di Napoli Federico II

Sorption Enhanced Gasification (SEG) represents a promising technology enabling to couple a CO_2 sequestration technique with the production of hydrogen-rich syngas by removing CO_2 from the reaction environment. Ca-based sorbents allow the capture of CO_2, being applied in the well-known â€œcalcium loopingâ€� post-combustion technique, typically carried out in Dual Interconnected Fluidized Beds (DIFB).

Six commercial limestones were investigated in terms of CO_2 capture capacity and attrition/fragmentation tendency under simulated SEG conditions and over repeated cycling typical of a looping process. A series of tests were carried out in a lab-scale DIFB system, each of which involving ten complete cycles of calcination/carbonation. Calcination was performed at 850 Â°C with a stream of 10% CO_2 in air (simulated oxidizing conditions of a combustor-calciner). As for the carbonation, a range of temperatures from 600 Â°C to 700 Â°C was investigated, which is the possible operation window of SEG found in the literature considering suitable gasification temperatures. Carbonation was carried out fluidizing the reactor with 10% CO_2 (balance nitrogen, to reproduce reducing conditions of a gasifier-carbonator). CO_2 capture performance during each carbonation stage was calculated by monitoring the CO_2 concentration at the exhaust. The sorbent attrition rate was evaluated based on the fraction of fines elutriated and collected in filters during each stage of a test. Exhaust samples, sieve-analyzed, were characterized obtaining the particle size distribution to evaluate the in-bed sorbents fragmentation extent. Moreover, the impact fragmentation extent was additionally evaluated by means of an *ex-situ* apparatus.

The results showed that, despite the similar chemical composition, the six sorbents behave differently. Overall, the attrition/fragmentation tendency is predicted to be relatively limited under simulated DIFB-SEG conditions, leading to conclude that the sorbent make up rate would mostly depend on the decay of CO_2 capture capacity due to sintering.

The Composition of Fluidised Bed Inventory Influences the Performance of Limestone Sorbent in Thermochemical Energy Storage Based on Calcium Looping

Authors

Di Lauro, F. - Presenter

Tregambi, C., Università degli Studi di Napoli Federico II

Montagnaro, F., Università degli Studi di Napoli Federico II

Molignano, L.

Salatino, P., Università degli Studi di Napoli Federico II

Solimene, R., Consiglio Nazionale delle Ricerche

Concentrated Solar Power (CSP) integrated with ThermoChemical Energy Storage (TCES) systems represents a promising technology to obtain an effective exploitation of solar energy and overcome its intrinsic intermittency. Since solid-gas reactions are favoured in TCES processes, fluidised bed reactors appear to be excellent particle receptors thanks to the good mass/heat transfer coefficients ensured by FB fluid-dynamics.

An interest reversible reaction for the above-mentioned purpose is represented by the calcination/carbonation of limestone seen as Ca-based sorbent thanks to the high reaction enthalpy (Î"H_rÂ°=178 kJ/mol), and low-cost of the raw material. Conversely, this process is penalised by the rapid deactivation of the limestone induced by loss of porosity following thermochemical sintering, pore plugging and low absorption of solar radiation, due to the high reflectance of limestone.

In this research work experiments were carried out in a lab-scale fluidised bed reactor, electrically heated. An Italian limestone was present in the bed alone, or in combination with silica sand/silicon carbide. SiC was chosen as representative of a high-absorbance bed material.

Calcination/carbonation cycles were carried out under operating conditions resembling the â€œclosed-loopâ€� scheme (calcination at 950Â°C, carbonation at 850Â°C, fluidising atmosphere composed by pure CO_2). Carbonation degree, particle size distribution and bulk density were measured as cycles progressed, together with the application of a model equation to relate carbonation degree to the number of cycles.

Mutual relationships between nature of bed material and possible interactions, degree of CaO carbonation, generation of fragments and changes in particle density were critically discussed. The possibility of separate bed particles at different carbonation degree by segregation patterns is also investigated. To this end, the segregation behaviour of binary mixtures of limestone and lime particles was scrutinized through tests in a fluidisation column equipped with a capacitive probe able to measure the composition of bed solids along the bed height.

The Role of Cohesion in the Formation of a Raceway of a Blast Furnace

Authors

Kuipers, S. - Presenter, Eindhoven University of Technology

Deen, N., Eindhoven University of Technology

Tang, Y., Eindhoven University of Technology

Huang, C. C., Eindhoven University of Technology

Blast-furnaces (BFs) are the workhorses of the modern steel making industry. The annual output of the steel-industry amounts to 1.9 Gt, which releases around 650 Mt CO_2 or 7% of the global annual emissions. The modern BF process makes use of lateral injection of a mixture of pulverized coal and hot air by circular tubes at high velocity, which results in turbulent flow and the formation of a void region, called raceway. Understanding raceway formation is of great importance for the BF performance, including the process efficiency as well as the emission reduction.

The raceway is often characterized by the morphology and dimensions. However these characteristics are extremely difficult to obtained experimentally under actual process conditions. Thus, numerical modelling using computational fluid dynamics (CFD) is an alternative. The contribution of this work is the consideration of cohesive particle-particle interaction in the CFD model to investigate its impact on the raceway formation. The modelled cohesive forces correspond with particle agglomeration (sintering). Simulations are performed using the CFD-DEM approach with incorporation of a simplified JKR model accounting for particle cohesive forces. Following verification with literature data the model was used to quantify the effects of cohesive particle-particle interaction and raceway characteristics, where we observed increased bed resistance of ca. 5%. This resulted in a collapse of the raceway at lower blast velocity. At higher blast velocity fluidization and gas leakage was suppressed, resulting in more uniform flow conditions.

Deep Learning for Drag Force Modelling in Dilute, Poly-Dispersed Particle-Laden Flows with Irregular-Shaped Particles

Authors

Fan, L. S. - Presenter, Ohio State University

Hwang, S., The Ohio State University

Pan, J., Ohio State University

This study employs machine learning-based approaches to develop a drag force model for particles of any irregular-shapes in incompressible gas-solid flows. The irregularity of the particle shape is described using the spherical harmonic model.
The gas-solid fluidization behavior represented bysparse, poly-dispersed particle-laden flows at low to intermediate particle Reynolds numbers is describedthrough in-house particle-resolved direct numerical simulations (PR-DNS). We utilize the PR-DNS to obtain drag force coefficients and flow fields of singleparticles. A variational auto-encoder model is appliedto obtain latent vectors to represent the geometrical features of the particles, with artificial neural networks (ANN) developed to predict drag force coefficients and flow fields of a single particle system. This study then applies a pairwise interaction extended point-particle (PIEP) model to obtain the drag coefficients of a single particle in multi-particle systems by assumingthe flow fields of individual neighboring particles can be linearly superposed over those of the single particle in consideration. The PIEP and ANN results show moderate correlation and accuracy based on the PR-DNS results. With the PIEP and ANN methods, this study provides a steady drag force model that does not require heavy data collection for the gas-solid fluidization systems with irregular-shaped particles.

Pressure Fluctuations in a Gas-Solid Fluidized Bed at Temperatures up to 1650°C

Authors

Bai, D. - Presenter, ReCommunity Inc

Fu, L., Shenyang University of Chemical Technology

Zhang, Q., Shenyang University of Chemical Technology

Xu, G., Shenyang University of Chemical

We, at the Shenyang University of Chemical Technology, have recently generated strong interest in transforming some ultra-high temperature solid-state reaction processes that have been taking place in traditional shaft/rotary kilns into fluidized bed operations, aiming to lower reaction temperatures and increase reaction rates, thereby cutting energy consumption and carbon emissions. For the successful development of energy-saving and environmental-friendly alternative technologies, it is significantly important to understand fluidization behavior at temperatures above 1000Â°C, but previous investigations at such high temperatures are scarce. In general, the operation of fluidized beds at such ultra-high temperatures is limited by defluidization caused by the agglomeration tendency of bed particles. For this reason, we conducted experiments in a laboratory fluidized bed of 30 mm diameter at temperatures from 800 to 1650â„ƒ with corundum and magnesite oxide particles. The pressure fluctuation signals are recorded at different superficial gas velocities and bed temperatures. Statistical and spectral analyses are performed to characterize high-temperature fluidization behavior by the corresponding minimum fluidization velocity, standard deviation, mean pressure amplitude, power spectrum, and dominant frequency. Great efforts are made to determine conditions at which stable fluidization or defluidization can be attained. The results show that we can achieve and maintain stable fluidization at ultra-high temperatures with some technical and experimental measures. The experimental results demonstrate that fluidization behavior at high temperatures is unique and cannot be fully predicted by the experience and theory applicable to low-temperature fluidization.

The Mechanism behind Vibro-Assisted Fluidization of Cohesive Micro-Silica

Authors

Kamphorst, R. - Presenter, Delft University of Technology

The mechanism behind vibro-assisted fluidization of cohesive micro-silica

Rens Kamphorst,Kaiqiao Wu, P. Christian van der Sande, Evert C. Wagner, M. Kristen David, Gabrie M.H. Meesters, J. Ruud van Ommen

In sectors such as food, pharma and catalyst manufacturing we see an increasing interest of applying fluidization to particles in the Geldart C range. When it comes to these particles, fluidization becomes challenging due to strong attractive inter-particle forces. This results in channel formation and agglomeration, both of which lead to a reduction in gas-solid contact. To fluidize cohesive particles, assistance methods can be employed, which introduce additional manipulation. For most assistance methods, studies have concluded that the mechanism by which fluidization was improved, is by breaking up agglomerates. We hypothesized that it is not the agglomerates, but the channeling that is targeted by these assistance methods.

In this study, we used X-ray imaging to study channel formation and collapse within a fluidized bed of cohesive microparticles. Pressure drop, bed height and off-line agglomerate size measurements were performed as well, to decouple the effect of channel disruption and breaking of agglomerates.

It was observed that channeling is the main cause of poor fluidization of the micro-silica, resulting in poor gas-solid contact and little internal mixing. Improvement in fluidization upon starting the mechanical vibration was achieved by disrupting the channels. Agglomerate sizes were found to not significantly change during experiments.

CFD-DEM Evaluation of Rod-like Particles Size Effect on the Mixing of Fluidized BED Containing Spheres and Rod-like Particles

Authors

Chaouki, J. - Presenter, Ecole Polytechnique Montreal

Kazemi, S., University of Tehran

Zarghami, R., University of Tehran

Mostoufi, N., Uniersity of Tehran

Sotudeh-Gharebagh, R., Uniersity of Tehran

Fuidization of particles can be used for a variety of purposes, including biomass gasification, drying applications, food processing, and pharmaceutics production. In practice, spherical particles are usually added to the bed to improve the fluidization quality of non-spherical particles. Understanding the particles mixing behavior in fluidized bed is essential in this case due to its role in process design improvement. Coupling of computational fluid dynamic-discrete element method (CFD-DEM) is used to investigate particles mixing in this study. Non-spherical particles are modeled by the multi sphere method. This work aims to investigate the effect of rod sizes on binary fluidization of spheres and rods. Also, the influence of the rod volume fraction and gas velocity on binary mixing are studied. The results illustrate that increasing the rod volume fraction and rods size increase the minimum fluidization velocity. Fluidized bed mixing becomes faster by reducing the rod-like particle size. Similar trend was observed by decreasing the fraction of rod-like particles. As a result, the addition of spherical particles and approaching particle shape to spherical leads to improvement of the fluidization quality and particle mixing. Furthermore, particle mixing improves by increasing the gas velocity in all investigated cases.

Experimental and Numerical Investigation of the Motion of an Object Immersed in a Fluidized Bed of Fine Particles

Authors

Ansart, R. - Presenter, Université de Toulouse, CNRS-Toulouse

Fotovat, F., Sharif University of Technology

Benguigui, W., Fluid Mechanics, Energy and Environment Department, EDF R&D

Lavieville, J., Fluid Mechanics, Energy and Environment Department, EDF R&D

Simonin, O., Institut de Mécanique des Fluides de Toulouse

Chaouki, J., Ecole Polytechnique Montreal

Characterizing the motion of a very large particle, called *object*, in a fluidized bed of fine particles is essential to understand the hydrodynamic phenomena occurring in gas-solid fluidized beds involving biomass, solid waste, or agglomerates. To this end, the position of a single spherical polymeric object immersed in a cylindrical fluidized bed of fine and coarse sand was tracked for a few hours by the radioactive particle tracking (RPT) technique. For each bed material, the experiments were carried out at two excess gas velocities, i.e. U_e = 0.25 m/s and U_e = 0.50 m/s. The analysis of the RPT data showed that the object rise takes place principally in the drift of the bubbles. Moreover, the mean object rise velocity to the mean bubble velocity was constant (~ 0.22), regardless of the fluidization medium and the tested excess gas velocity. It was also noticed that the residence time of the object in the emulsion phase depends on the fluidization medium and this parameter influences the object rise frequency. To understand the physical phenomena involved at meso-scale and to guide/validate the mathematical modeling, the experimental cases were simulated by an Euler-nfluid approach based on the kinetic theory of granular flow representing both gas and sand phases. To track accurately the motion of the object, a discrete forcing method based on a porous medium approach was proposed which could strictly ensure the conservation laws at the close vicinity of the interface. The interface is approximated as a plane in each cut-cell. The domain contains the object structure, which is considered a real part of the calculation domain. The use of an immersed boundary method to perform fluid-structure interaction is powerful as it allows large or slight displacement without any dependency on the mesh.

Fundamentals Keynote-Investigation of Wet Particle Contact Mechanics

Authors

Holland, D. J. - Presenter, University of Canterbury

Many industrial applications and natural phenomena consist of mixtures of granular materials and liquids. The inclusion of the liquid forces creates complex contact dynamics. Our understanding of the effect of the liquid is still limited and accurate modelling of these systems is a challenge. In this study, we consider particles coated in thin layers of viscous fluids such that the capillary number is high and viscous forces dominate.

Various models have been proposed to characterise the contact mechanics for particles coated in viscous liquids. Although all based on viscous flow, these models introduce quite different approximations to yield a computationally viable model. Here we present a detailed comparison of the different models with experimental measurements. High-speed imaging and a novel electrical conduction measurement are used to experimentally investigate the particle collision dynamics for linear and oblique collisions. We find that the models can predict the collision outcomes accurately, however they require the inclusion of an empirically fitted, non-physical parameter. It appears that this parameter arises from an overestimate of the viscous forces estimated by the models. The talk concludes with some suggestions regarding how the viscous force models could be improved to better replicate the experimental results.

Float-Sink of Large Objects in a Gas-Solid Fluidized Bed Different from That in a Liquid

Authors

Oshitani, J. - Presenter, Okayama University of Science

Tsuji, T., Osaka University

Harada, S., Hokkaido University

The gas-solid fluidized bed has liquid-like properties. Large objects float or sink in the fluidized bed based on density difference like in a liquid. Dry separation using the float-sink is in great demand as an alternative to the commonly used wet separations which have various environmental issues. The fundamental studies and applied research in the field of mineral processing and waste treatment has been actively carried out. The float-sink has been discussed based on Archimedes principle, considering the density difference between the fluidized bed and the objects. Although Archimedes principle properly works to explain most of the float-sink, we observed some float-sink phenomena deviated from Archimedes principle. Here we report the three anomalous float-sink phenomena. 1) A smaller density sphere sinks deeper than a larger density sphere when the air velocity is just below the minimum fluidization velocity (u_0/u_{mf} = 0.95) and the sphere density is close to the bulk density of the powder bed (ï�$_{sphere}$/ï�$_{bulk}$ â‰ˆ 1). The larger density sphere sinks instantly without air bubbles, whereas the smaller density sphere sinks slowly with air bubbles around it. 2) Spheres tend to sink with increasing u_0/u_{mf} from 2 to 4 due to the bed expansion and the decrease of the apparent fluidized bed density, whereas the spheres sinking becomes unstable at $u_0/u_{mf} > 4$. The float-sink depends on the fluidized powder bed height. 3) When the sphere density is just below the apparent fluidized bed density, the sphere floats if only one sphere is put onto the fluidized bed, whereas the spheres sink if many spheres are put onto the fluidized bed. The possible mechanisms will be discussed with considering the local fluidization around the sphere, the fluid force affected by the bubbles motion and the fluidized powder bed motion, many-body effect in the fluidized bed and so on.

Analysis of Horizontal Solids Convection and Dispersion in Fluidized Beds

Authors

Farha, M. - Presenter, Chalmers University of Technology

Guío-Pérez, D. C., Universidad Nacional de Colombia

Johnsson, F., Chalmers University of Technology

Pallarès, D., Chalmers University of Technology

The horizontal transport of solids in a fluidized bed is typically described by means of a dispersive process. A solids crossflow, however, also involves a macroscopic convective component, which needs to be considered. The latter applies to systems consisting of interconnected fluidized beds, which are expected to form the basis for various applications in the future energy system. Understanding the interplay between dispersive and convective solids mixing in the lateral direction is key to determining the solids residence time distributions and, thereby, the closure of the mass balance in interconnected fluidized bed reactors with a horizontal solids crossflow. This work aims at investigating the lateral transport mechanisms for solids in a bubbling fluidized bed with such horizontal solids cross flow.

Lateral solid dispersion induced by the bubble flow is quantified and compared with a net horizontal solids convection forced onto the bed, along with the interaction between dispersion and convective processes. The experiments are conducted in a set-up designed to force a horizontal flow of bubbling fluidized solids around a closed horizontal loop. The loop contains a horizontal straight channel through which the concentration of a ferromagnetic solids tracer injected upstream is measured non-intrusively (at 100 Hz) by means of an impedance coil at four different positions along the channel. The set-up is designed in accordance with Glicksman's complete set of fluid-dynamic scaling laws, allowing measurement scale-up to large-scale hot conditions. The measured values are used to fit a mathematical model of a one-dimensional transient convection-dispersion process, thereby obtaining the lateral solids dispersion coefficient. Three operational parameters were examined: bed height, fluidization velocity, and lateral solids convection.

Instabilities in an Underflow Standpipe Operating in a Bubble-Upflow Regime

Authors

Karri, S. B. R. - Presenter, Particulate Solid Research, Inc. (PSRI)

Findlay, J., PSRI

Du, B., ExxonMobil Research and Engineering

Harandi, M., ExxonMobil Research and Engineering

Testing was conducted to understand the operation of an underflow, bottom-constrained vertical 25-cm-diameter standpipe operating in a bubble upflow regime with Geldart Group A FCC equilibrium catalyst. Standpipe operation was studied over a range of solids mass fluxes, amount of external aerations at the bottom of the standpipe with slide valve controlling the catalyst circulation rates. The effects of both aeration amount and catalyst flux on standpipe operation in a bubble upflow regime were investigated. Test results yielded the following findings:

Demonstrated stable operation of the standpipe over a range of catalyst fluxes up to 60 kg/s-m² and an superficial gas velocity of 0.04 to 0.15 m/s in the standpipe. Under these conditions, the bubbles tend to flow upward against downflowing catalyst,

Stable operation was only achieved when the catalyst with fines content less than 44 microns, was increased to at least 7 to 8% by weight in the circulating catalyst. Increasing flow instabilities and catalyst flow problems were observed as the fines content of the catalyst decreased below these levels,

Helium tracer gas testing indicated that gas by-passing is occurring in the 6 m tall standpipe, even at the 8% fines content. Research at PSRI have shown that the pressure ratio across a fluidized bed has a direct impact on the probability that gas by-passing will occur in deep beds. Gas bypassing in the standpipe was the root cause of instabilities in standpipe operation,

For low fines catalyst, catalyst flow became unstable at very low solids flux in the standpipe, when the bottom aeration gas flow rate was increased to provide superficial gas velocity exceeding 0.08 m/s in the standpipe. The differential pressures (â^†Ps) above and below the air sparger increased sharply and remained high for 10 to 20 seconds before returning to near normal for a short period.

3D Numerical Simulation of the Heat Transfer in a Horizontal Tube Bundle in a Vertical Channel

Authors

Corcoles Tendero, J. I. - Presenter

Díaz-Heras, M., Castilla - La Mancha University

Domínguez, P.

Almendros Ibáñez, J. A., Universidad de Castilla-La Mancha

In fluidized bed technology, one of the most common processes is the heat transfer between gas and horizontal tube bundle. Hence, immersed surfaces into the fluidized bed are useful to extract heat from it for several industrial processes. To study the influence of the tube bundle distribution is needed to show hydrodynamic analysis, mainly focused on the bubble size and velocity, as well as particle mixture rates.

In the present study, a 3-D numerical simulation is implemented in Barracuda Virtual Reactor (v. 22.1) to perform a hydrodynamic and heat transfer analysis of a fluidized bed heat exchanger. It is based on the experimental study carried out by Moawed et al. (International Journal of Chemical Reactor Engineering, 2010), who used a bundle of tubes in a vertical channel. In line and staggered tubes arrangement have been implemented.

Local behavior of the heat transfer coefficient and particle volume fraction have been studied at different angular positions of a central tube, to properly analyze the relationship between hydrodynamics and heat transfer phenomena. In both tube bundle distribution, the lowest heat transfer coefficients values were obtained at the top region of the tube, which is related to the fact that particles are at stagnant and hot particles close to the surface are not being replaced by new cold particles. This behavior is different at the bottom and sides, where particles are continuously being removed, obtaining low particle volume fraction values and high heat transfer coefficients.

In both cases, the tube bundle limits the bubble coalescence and maintain a low bubble size along the bed. About in line tube bundle, the stagnation region size located at the top is slightly higher than at the staggered tube arrangement, mainly related to the zigzag shape of the air flow circulating around the tube.

The Influence of Fluidising Gas Density and Viscosity on Segregation in a Cylindrical Fluidised Bed Separator

Authors

Saw, W. L. - Presenter, The University of Adelaide

Ping, S., The University of Adelaide

Nathan, G. J., The University of Adelaide

Guo, P., The University of Adelaide

This study investigates the segregation behaviour of a fluidised bed separator by assessing the effects of density and viscosity of the fluidising agent. Segregation can be desirable for some applications, such as for the continuous removal of one type of product or for classification of particles. In particular, their impact on the bed composition, bubble formation, pressure fluctuation, and superficial gas velocity are discussed. The particle distribution profile along with the bed height and pressure drop are measured, together with the pressure fluctuation profile and mixing index. The results show that varying the ratio of solid to gas density has a negligible impact on segregation for superficial gas velocities between 1.2 (minimum fluidisation velocity) and 3.5 . The gas viscosity has a significant impact on segregation for a superficial gas velocity of between 1.2 and 1.5 , but its impact reduces as the superficial gas velocity and bed composition are increased. The high gas viscosity case was found to lead a suppression of the bubble formation and pressure fluctuation, thus resulting in low segregation. Pressure fluctuations further demonstrate the inhibitory effect of high gas viscosity on bubble formation, and the experiments discovered that high gas viscosity promotes fluidised bed stability. It is noteworthy that although the present experiments were carried out in a lab-scale fluidised bed at ambient temperature, the particle density ratio and the particle to gas density ratio are based on the dimensionless scaling numbers that can provide a guide to the segregation in the gasification system using sand and char. The large range of gas viscosities examined in the experiments also can assist with the control of high viscosity steam gasification system.

Heat Transfer in a 3D Dynamically Structured Fluidized Bed

Authors

Jiang, S. - Presenter

Wu, K., Delft University of Technology

Coppens, M. O., University College London

Inspired by the origin of ordered structures generated by periodically oscillating fluid flows in nature, such as the sandy ripples on beaches and desert floors, it has been proposed to apply an oscillating gas flow at a state around the minimum fluidization velocity to create a â€œdynamically structured fluidized bedâ€, which manipulates the gas-solid drag flow periodically in time. The pulsed flow introduces order, and the system reaches a specific fluidization state between fixed and fluidized bed. In quasi-2D flat or annular fluidized beds, bubbles rise and form a hexagonal array with bubbles spaced at certain distances, which is different from the chaotic bubble flows in conventional fluidization. This special fluidization state provides opportunities for process intensification. Previous research has investigated the hydrodynamics in dynamically structured fluidized beds and identified the operation of a structured bubbling flow with bubble size and pattern wavelength that could be tuned by changing the oscillation frequency. However, the influences of dynamical structuring on transfer coefficients are still poorly understood.

An essential step towards utilizing dynamically structured fluidized beds for, e.g., heterogeneously catalysed reactions or drying, is the development of a heat transfer model that considers the unusual characteristics of dynamically structured fluidized bed, lying in between traditional fluidized and fixed beds. In this work, the heat transfer characteristics within a 3D annular pulsed fluidized bed were first studied numerically, and subsequently compared with the unstructured conventional bubbling fluidized bed. By using a thermocouple, the instantaneous temperatures were measured at various locations, and the simulation results showed good agreement with experimental results. The relationship between bubble patterns and temperature distribution was explored, and the effects of pulsed flow on local and overall heat transfer coefficients were also investigated.

CFD-DEM Coupled Simulation for Wet Particles Drying Process in a Fluidized Bed

Authors

Wang, H. Sr. - Presenter, Institute of Engineering Thermophysics

Yang, W., The University of Manchester

Ma, Z., Chinese Academy of Sciences

Tu, Q., Chinese Academy of Sciences

Ocone, R., Heriot Watt University

Wet particle drying is complex due to the variation of particle moisture which has big effect on the gas-solid flow dynamic characteristics as well as mass and heat transfer between particles and particles, particles and flow. CFD-DEM simulation were employed to investigate the main characteristics in a fluidized bed drying process. Due to the computational capability restriction and computational time limit, 2D model simulation were first employed for the whole drying and then 3D model simulation was employed at certain moisture contents for more accurate flow dynamics. The simulation was based on the coupled CFD-DEM approach. The objective of this research is to further investigate the details of the wet particles flow characteristics, the heat transfer and mass transfer process of the particle drying process in a fluidized bed dryer and the effect of the particle number and the changing liquid bridge force on the particle drying process. The simulated cross-sectional particle volume fraction shows core-annular structure and is verified by the ECT measurement. An internal circulation is formed in the height direction, and the mass and heat transfer mainly happen at the lower part of the dryer and at the contact region between the bubbles and the solid phase. The results predicted by the CFD-DEM model and the TFM model are compared.

Startup and Control Concepts for Pressurized Chemical Looping for a Plug Flow with Internal Recirculation Reactor

Authors

Bond, N. - Presenter

Champagne, S., CanmetENERGY

Alain, A., CanmetENERGY

Hughes, R. W.

McIntyre, C., Hatch Engineering Ltd

Pressurized chemical looping combustion (PCL) is a carbon capture technology that produces high-temperature heat with inherent separation of CO_2. It employs two fluidized beds to avoid direct contact of the combustion air and fuel, instead using a solid metal oxide oxygen carrier that transports the oxygen from the air reactor to the fuel reactor, avoiding dilution of the produced CO_2 with N_2. In collaboration with Hatch Ltd, CanmetENERGY is building a 600 kW_{th} PCL pilot plant using Hatchâ€™s novel plug flow with internal recirculation reactor (PFIR) concept. This concept houses both the air and fuel reactors within a single vessel, making the technology more compact, reducing capital costs, and eliminating reliability and operability issues typically associated with external solids transfer legs in traditional chemical looping arrangements. This novel reactor , while advantageous in many aspects, introduces new process design requirements associated with: (1) control of circulation between reactors, (2) control of product purity from the air and fuel reactors, (3) startup of the process and (4) pressure control in the reactors.

CanmetENERGY is developing startup and process control strategies for the pilot plant that can be tested and then implemented at commercial scale. The startup strategy is a multi-stage approach that includes pre-heat and fluidization of the air reactor with a natural gas burner, and use of a recycled gas stream to the fuel reactor to fluidize and drive circulation of the cold fuel reactor bed particles. Initiation of solids circulation between reactors is an essential part of startup to transfer sensible heat from the air reactor to the fuel reactor. Transition from startup to chemical looping combustion must also be carefully balanced to ensure that fluidization, heat transfer between the reactors, reaction stoichiometry and reactor pressures are all balanced.

Impact of Process Conditions on Fluidized Bed Drying: Experiments and Model Predictions

Authors

Leung, C. - Presenter, Rutgers University

Glasser, B., Rutgers University

Shapley, N. C., Rutgers, the State University of New Jersey

Langrish, T. A. G., The University of Sydney

Fluidized bed drying is an essential unit operation in the manufacturing of supported catalysts. Heated gas flows through a bed of wet catalyst supports which promotes fluidization, allowing for high rates of heat and mass transfer between the gas and solids. Although many studies have been conducted in fluidized bed dryers, general prediction equations for fluidized bed drying of catalysts are lacking, as drying depends on the fluidization conditions and support properties. Several phenomena, including heat and mass transfer, drying and fluidization, makes it difficult to choose the operating conditions to optimize the drying process. In general, one would like to have sufficient understanding of the process so that laboratory scale experiments can be used to predict behavior at the pilot and manufacturing scales.

Experiments were carried out in a Glatt GPCG-1 fluidized bed dryer. Water impregnated catalyst supports were used as the experimental material. The change in moisture content of the bed over time was measured along with the bed temperature and the rate of drying was calculated. Different bed loading, inlet air temperature, initial moisture content, and gas flowrate were tested. Using experimental reference drying curves, model equations for predicting the moisture content of the bed were implemented. For a large range of process conditions, we found that the model worked best when the drying rate was roughly constant, which correlates to the unhindered or constant rate drying period. However, the model was still able to make reasonably good predictions, even when drying started in the hindered or falling rate drying period. For high bed loadings, where the number of transfer units was large, the experiments and model showed the best agreement. This work provides quantitative findings on how process conditions affect the rate of drying for a bed of catalyst supports with corresponding prediction equations.

The Techno-Economics of Electrification of Heat and Power Plants with Thermochemical Energy Storage - a Swedish Case Study

Monday, May 22, 2023 - 4:51pm to 4:53pm

Authors

Martinez Castilla, G. - Presenter, Chalmers University of Technology

Guío-Pérez, D. C., Universidad Nacional de Colombia

Cortés Romea, J., Chalmers University of Technology

Toktarova, A.

Pallarès, D., Chalmers University of Technology

Johnsson, F., Chalmers University of Technology

Thunman, H., Chalmers University of Technology

Achieving high penetration levels of renewable electricity is crucial in order to decarbonize the energy system. However, a large share of renewable electricity will be in the form of variable renewable electricity (VRE) technologies, i.e., foremost wind and solar power, which rely on efficient strategies to include flexibility in the energy system, such as the introduction of energy storage in order to maintain the value of VRE. Thermochemical energy storage (TCES) systems offer a possibility for long-term storage at ambient conditions and with energy densities of the same order of magnitude as currently used carbon-based-fuels.

This work proposes and assesses the use of heat and electricity generation with incorporation of a TCES using high temperature redox reactions in large-scale fluidized bed reactors. Sweden is used as a case study and therefore- considering the presence of iron-ore mining industry and the development of hydrogen-based steel production - iron oxide is chosen as the investigated system. The possibility for retrofitting existing biomass-fueled fluidized bed combustion (FBC) plants is assumed, since Sweden has a relatively large fleet of FBC plants in operation that will need to provide more flexibility due to increased share of VRE and surge in biomass prices. The reduction step utilizes hydrogen produced via electrolysis during periods of competitive electricity prices and the oxidation reactor (retrofitted boiler) dispatches heat and electricity to meet the demand. The work includes i) the computation of energy and mass balances of the process, ii) the construction of a plant investment cost function and iii) a cost-optimal sizing of the process for three different scenarios of the development of the Swedish electricity system. The results of the work indicate the ability of the proposed scheme to generate heat and electricity at a preliminary cost (including CAPEX and OPEX) of 20-70 eur/MWh_{heat} after the retrofit.

Electrostatics Charging Behavior of a Metallocene Catalyst in Gas-Solid Fluidization of Polyethylene

Authors

Isaac Nimvari, M. - Presenter

Mehrani, P., University of Ottawa

Sowinski, A., University of Ottawa

Contact electrification is a surface phenomenon in which materials acquire electrostatic charge after contact and separation, potentially resulting in undesired occurrences of solids agglomeration and fouling. For instance, in the commercial gas-phase polyethylene (PE) fluidized bed reactors electrostatics could cause reactor wall fouling and sheeting, in turn necessitating reactor shut down. In contact charging, the amount of transferred charge and the generated polarity largely depend on solids surface chemistry. In this study, in relation to the polyethylene process, for the first time the role of presence of a commonly used catalyst, and the charging behaviour of components present on the catalyst surface were investigated in an atmospheric gas-solid fluidized bed under cold flow conditions.

A 0.1 m in diameter stainless-steel fluidized bed was used to fluidize a linear low density polyethylene (LLDPE) resin received from commercial reactors. A metallocene catalyst in its active and inactive form was tested along with a bottom-up approach in which the catalyst base and the catalyst activator were examined separately. Fluidization using nitrogen at 1.5 minimum fluidization velocity was carried out for a period of 1 hr where the powders were added to the bed at the onset of fluidization. The column contained two Faraday cups; one below the distributor plate to measure the particles charge inside the column (â€œ*Bulkâ€*�*)* or adhered on the wall â€œ*Fouling*â€�, and one above the freeboard to measure the entrained particles charge during fluidization.

Results showed that catalyst base alone induced a strong negative polarity into the bed of polyethylene and increased the amount of *Fouling*. Adding a co-catalyst changed the *Bulk* polarity and thus the amount of *Fouling* declined. The active catalyst, however, did not change the polarity or the magnitude of *Bulk* charge, in turn did not result in significant changes in the amount of *Fouling.*

Fluidized Bed Surface Finishing of Metal Objects Produced By Additive Manufacturing

Authors

Troiano, M. - Presenter, Università degli Studi di Napoli Federico II

el Hassanin, A., Università degli Studi di Napoli Federico II

Silvestri, A. T., Università degli Studi di Napoli Federico II

Scherillo, F., Università degli Studi di Napoli Federico II

Scala, F., Istituto di Ricerche sulla Combustione, Consiglio Nazionale delle Ricerche

Solimene, R., Consiglio Nazionale delle Ricerche

Squillace, A., Università degli Studi di Napoli Federico II

Salatino, P., Università degli Studi di Napoli Federico II

Fluidized Bed Machining (FBM) of additively manufactured objects is being considered as a viable post-processing method for object finishing and improvement of surface texture. As-built objects are characterized by high surface roughness, thus, post-processing is required. Conventional finishing methods are limited to objects with low geometrical complexity, while in principle, fluidized bed finishing can be suitable for complex geometries, thanks to the mobility of fluidized particles.

The study addresses a comprehensive assessment of FBM technology aimed at the surface finishing of square flat metal specimens produced by Laser-Powder Bed Fusion method. The experimental campaign aimed at characterizing the effectiveness of two operational modes: FBM of stationary specimens immersed into a fluidized bed of abrasive material; FBM of specimens immersed in the fluidized bed under controlled rotational motion. The effect of different abrasive materials, processing time and tilt angle of the specimen was investigated. Surface modifications were analyzed in terms of mean surface roughness, skewness and primary profile root mean square slope, as well as by Scanning Electron Microscopy and monitoring of time-resolved weight loss. The results showed moderate smoothing of the sample surface under stationary sample configuration, while good results were obtained for rotation-assisted FBM tests. A maximum reduction of the surface roughness of 12% and 67% of the initial value was measured for stationary and rotation-assisted FBM, respectively. An appreciable influence of the bed material and of the tilt angle was observed. Steel particles were the most effective bed material for the finishing process, with particle density overtaking hardness as the key particle property. The optimal tilt angle was the one that maximized sliding and shear. The experimental results are consistent with a mechanistic framework according to which surface finishing of the specimens is driven by surface shear forces dominated by inertial stresses in the particulate phase.

Using a Heat Balance to Determine Solid Circulation Rate in a Cold Flow Plug Flow Reactor with Internal Recirculation

Authors

Alain, A. - Presenter, CanmetENERGY

Bond, N.

Champagne, S., CanmetENERGY

Hughes, R. W.

Macchi, A., University of Ottawa

McIntyre, C., Hatch Engineering Ltd

Francey, S., Hatch Ltd

A novel system for pressurized chemical looping (PCL) called a plug flow with internal recirculation (PFIR) reactor is being developed at CanmetENERGY in collaboration with Hatch Ltd. The PFIR is a novel, compact, annular fluidized bed configuration incorporating directional jets that circulate solids between fluidized bed segments in a single vessel. This PCL reactor design eliminates the need for two separate fluidized bed reactors and the associated external solids transfer equipment. It allows for relatively high oxygen carrier circulation rates between the air and fuel reactors, which is an important design challenge for PCL systems. Demonstrating that high oxygen carrier solid circulation rates can be achieved in a 0.61 m outer diameter cold flow PFIR system is a key milestone in confirming that PFIR reactors can be used for PCL. Studies on solid circulation rate in the PFIR are being used to better understand this unique reactor design and to validate a computational particle fluid dynamics (CPFD) model of the PFIR.

A technique using a heat balance to determine solid circulation rate has been investigated. Particles are heated *in situ* with a heater and the temperature is measured at the transition points between the reactor sections. Parameters that affect gas momentum will influence the solid circulation rate in the PFIR. For this reason, the effect of PFIR geometry and operating conditions, such as pressure at the bottom of the bed, on solid circulation rate is examined.

Rapid Stabilization of Aluminum Dross Ash By Fluidized Bed Steam Treatment

Authors

Noda, R. - Presenter, Gunma University

Togashi, H., Gunma University

Futawatari, A., JEMCO Co. Ltd.

Aluminum dross ash, a difficult-to-recycle waste generated in the aluminum industry, has been mostly disposed of. Aluminum dross ash contains aluminum nitride to form ammonia in the disposal, so that stabilization treatment has been needed to remove nitrogen. The most common stabilization treatment is a wet process in which aluminum dross ash is immersed in water to remove nitrogen as ammonium ion into the solution. While this process has low initial cost, it also has a problems like generation of a large amount of ammonia-containing waste water and long processing in several days order.

The authors have been proposed a new dry treatment for aluminum dross ash stabilization by fluidized bed steam processing. A laboratory scale fluidized bed was prepared to demonstrate the denitrification characteristics of actual aluminum dross ash in steam fluidization. In the experiments, aluminum dross ash was mixed with silica sand to achieve the smooth fluidization.

The effects of treatment temperature and steam concentration were confirmed at 110, 150, 170 and 200 °C and in 30, 40, 60 vol%. From the experiments with the superficial gas velocity of around 2 u_{mf}, it was found that reaction temperature of 150 °C and steam concentration of 40% were appropriate for nitrogen removal from aluminum dross ash. In the appropriate temperature and vapor pressure, the nitrogen removal rate gradually decreased and eventually became zero in the fixed bed, whereas the nitrogen removal was continued without a significant decline of the reaction rate in the fluidized bed.

Those phenomena were probably caused by aluminum oxide layer formed on the surface of aluminum nitride acting as an inhibitor of the reaction. In fluidized bed, the oxide layer could be updated by particle collision, so that the steam treatment by a fluidized bed is promising for aluminum dross stabilization.

Magnetic Particle Tracking for Granular Flows: A Semi-Algebraic Solution

Authors

Buist, K. - Presenter, Eindhoven University of Technology

Nijssen, T., Delft University of Technology

Magnetic Particle Tracking (MPT) is a non-invasive measurement technique often used to study dense granular flow, particularly for fluidization and rotating drum equipment. The basic principle of MPT relies on tracking the movement of a single magnetic tracer by means of measuring the magnetic field strength with a set of sensors. The difference between the sensor measurements and an estimate of the magnetic field, based on the position and orientation of the magnet can be solved by means of minimization techniques. MPT is therefore uniquely suited for the study of non-spherical particles, since it provides additional information on the orientation of the tracer. The performance of the localization is largely dependent on the signal-to-noise ratio and so far mostly relies on nonlinear optimization techniques. The definition of the magnetic field generated by a dipole is highly nonlinear and has five degrees of freedom. Here we present a semi-algebraic solution by decoupling the estimation of the position and orientation in separate algebraic solutions. The two estimates are mutually dependent, necessitating an iterative approach between the two. The main benefits of this new approach are the speedup and robustness of the algorithm, which are much higher than for the classical constrained nonlinear optimization techniques. It therefor allows both for fast screening of the initial data obtained by MPT and potentially could help screen sensor layout designs. Further improvements on this technique have been made with help of vectorization of the problem.

Effective Drag on Spheres, Rods, and Discs Immersed in a Fluidized Bed at Minimum Fluidization - Measurements and Modeling Work

Authors

Guío-Pérez, D. C. - Presenter, Universidad Nacional de Colombia

Köhler, A., Chalmers University of Technology

Nikku, M., LUT University

Myöhänen, K., 2) Lappeenranta-Lahti University of Technology

Pallarès, D., Chalmers University of Technology

With their excellent mass, momentum, and heat transfer properties gas-solid fluidized beds (FB) are used in a variety of applications, involving the presence of larger particles immersed in beds of smaller bulk solids (e.g., fuel particles in a sand/ash bed). It is known that the interaction between the large particles and the emulsion of bed material impacts the bed-to-particle mass and heat transfer rate, thus, the particle conversion. However, currently there is not a complete theory of the momentum exchange between the large particles and the emulsion of bulk solids, due to the complexity of the rheological behavior of the latter. Such information is vital for the development of reliable descriptions of the mixing mechanisms that the large particles exhibit in the bed, so that the design and scale-up of FB processes could be done more precisely.

Based on the hypothesis that gas-solid FBs (of Geldart-B-type particles) feature a non-Newtonian rheological behavior, the authors of this work have previously developed an experimental methodology to study the momentum exchange between immersed objects and gas-solids emulsions at minimum fluidization. By means of falling object experiments with magnetic particle tracking the instantaneous velocity can be measured, which allows the assessment of the shear stress on the immersed objects, and consequently the analysis of the rheological behavior of the emulsion.

Taking departure in the findings from the works done with spherical objects (of varying size and density) in fluidized emulsions of various bulk solids (of varying size, shape, and density) the present work aims to synthesize the current understanding and, through experimental data, extend it to non-spherical objects (rods and discs). Additionally, initial computational fluid dynamics (CFD) simulations are performed and compared with experimental results to evaluate the suitability of available numerical models. Based on this comparison, need for possible model development is discussed.

Enhanced Fluidization of Cohesive Powders

Authors

Wu, K. - Presenter, Delft University of Technology

Wagner, E. C., Delft University of Technology

Van Ommen, J. R., Delft University of Technology

M.H. Meesters, G., TU Delft

Because of their extremely high specific surface area, fine powders are of great interest for a wide range of applications in the pharmaceutical, chemical and food industries. Nevertheless, decent fluidization of fine powders is yet problematic due to their cohesive nature. At a steady gas flow, the interparticle forces promote the formation of agglomerates among powders, and give rise to persistent gas channels bypassing a significant amount of the inlet gas continuously. Such poor interphase contacts significantly depress overall mixing as well as mass and heat transfer [1]. Over the last decades, many studies adopted additional manipulations on gas or solids, including mechanical vibration, mechanical stirring, and gas pulsation, to alter the fluidization of fine powders [2]. Despite the augmented fluidization observed, the hydrodynamics created and underlying physics remain largely unexplored.

In this contribution, we experimentally compare the fluidization across different industrially relevant assistance, including mechanical vibration and gas pulsation. Fast X-ray imaging, as a non-intrusive measurement technique, provides direct monitoring of local hydrodynamic events and global fluidization development [3]. We also discuss the impact of varying vibration and pulsation conditions on hydrodynamic events such as powder segregation, bubbling and channel formation. The degree of enhancement can be assessed according to the comparison of assisted and conventional fluidization. These results help guide the design, implementation, and scale-up practices of relevant solid processing applications.

[1] J.R. van Ommen, J.M. Valverde, R. Pfeffer, Fluidization of nanopowders: a review, J. Nanopart. Res. 14 (2012) 737.

[2] V. Francia, K. Wu, M.-O. Coppens, Dynamically structured fluidization: Oscillating the gas flow and other opportunities to intensify gas-solid fluidized bed operation, Chem. Eng. Process. 159 (2021) 108143.

Experimental and Numerical Validations of Energy Storage By Sand in a Fluidized Bed

Authors

Liu, Z. - Presenter, IHI Corporation

Ishikawa, A., IHI Corporation

Zhang, H., Tsinghua Univercity

Hashiba, M., Technology Platform Center, Technology & Intelligence Integration

Zhang, Y., Tsinghua Univercity

Yang, W., Tsinghua Univercity

Yifei, Z., Tsinghua Univercity

The Variable Energy Resource, such as wind and solar power, is developing very rapidly. However, electricity generation by Variable Energy Resource is usually unsteady and cannot match the demand consumption of electricity. Therefore, large-scale, low-cost energy storge technology is required. Energy Storage by sand in a Fluidized Bed (ESFB) is proposed by IHI to satisfy this requirement.

Electrical heaters are set in bottom of riser of circulating fluidized bed. Sand is fed from the low-temperature tank to riser, heated by heater and simultaneously lifted to the top of riser, then stored in the high-temperature tank, when electricity is exceeded. The hot sand is fed from the high-temperature tank to heat exchanger and then go to the low-temperature tank, to generate steam and then generate electricity when it is required.

ESFB has the advantage of low cost, easy to scale-upï¼Œand high flexibility towards varying electricity demand/consumption. It also has the advantage of both high energy density and high safety.

The hot experiments of 10kW and 85kW have been carried out to validate the concept of ESFB. The numerical simulations have been made for optimizations. Both the experiments and simulations show that ESFB can work well.

The Effect of Clusters on the Heat and Mass Transfer in Fluidized Gas-Particle Flows

Authors

Schneiderbauer, S. - Presenter, Johannes Kepler University

Rauchenzauner, S., Johannes Kepler University

Gas-particle flows play a crucial role in modern industrial applications. In times of an intense requirement of enormous energy savings in industrial reactors, simulation and modelling is an important tool to asses those processes. Still, computational resources are limited and the time needed to simulate even a few seconds of process time in full-scale reactors is unfeasible for the industry. Although CFD is a potent tool, which can uncover the flow inside the reactors and newer methods, such as the Discreet Element Method (DEM) for particles, enable potent descriptions of the micro-scale flow, the multi-scale nature of gas-particle flows hinders the use of coarse numerical grids, which can speed-up the simulation time. Essentially, it was observed that the meso-scale flow structures, such as particle clusters and streamers, have a crucial influence on the macro-scale flow properties and not accounting for them leads to a considerable overestimation of the drag as well as heat and mass transfer in the system.

In particular, the mass transfer between the phases is also greatly affected by the clustering of the system. We identify the main properties in the species transport equation by application of spatial filters. We find that the species diffusion, which is a function of the particle Reynolds number, is severely overestimated if the particle clustering is not considered. In accordance to the better-known drift velocity correction to the resolved drag force and the newly established drift temperature correction to the heat transfer, we propose another drift scalar approach in order to qualify and quantify the correction to the resolved mass transfer in coarse-grid CFD simulations. In the framework of multi-phase turbulence modelling, the drift scalar is then expressed as a covariance between the solid volume fraction and the species concentration and modelled using a dynamic test-filtering approach.

PIV and Pressure Measurements for Fluidized Bed Model Validation: Building a Multi-Regime Reference Data Set

Authors

Porter, L. - Presenter, TU Bergakademie Freiberg

Richter, A., TU Bergakademie Freiberg

Gräbner, M., TU Bergakademie Freiberg

Zink, M., TU Bergakademie Freiberg

Fluidized beds are widely used in industry for various purposes, such as combustion, gasification, and catalysis. However, accurately modeling the complex fluid dynamics of fluidized beds remains a challenge, and model validation is critical for their reliable design and operation. In this study, we present a reference data set for bubbling fluidized bed (BFB) model validation based on particle image velocimetry (PIV) and pressure measurements.

The experiments were conducted in a lab-scale quasi-2D test rig comprising two different geometries. Two distinct particle size classes with polydisperse distributions were fluidized with air at different superficial velocities. Additional varied parameters include particle density and solids inventory. PIV was used to measure the velocity field of the particles, while pressure sensors at various locations record pressure data. The resulting data set includes pressure distribution and fluctuation, time-resolved velocity fields, and velocity frequency distributions for a range of operating conditions.

The data set will be made publicly available and is aimed to serve as a valuable resource for researchers and engineers to validate their BFB models under various operating conditions. It enables the evaluation of model accuracy for the bubbling fluidized bed regime and provides insight into the underlying hydrodynamics of the BFB system. In a next step to improve the data set bubble analysis will be conducted. Furthermore it is planned to expand the data set by extending the investigated regimes beyond bubbling regime to cover also turbulent regimes in an effort to build a true multi-regime validation data set.

Directly Irradiated Fluidized Bed Autothermal Reactor (DIFBAR): High-Temperature Operation

Authors

Padula, S. - Presenter, Federico II University

Troiano, M., Università degli Studi di Napoli Federico II

Tregambi, C., Università degli Studi di Napoli Federico II

Solimene, R., Consiglio Nazionale delle Ricerche

Salatino, P., Università degli Studi di Napoli Federico II

There is growing interest toward the application of Concentrating Solar Thermal (CST) technologies to endothermic processes, involving gas-solid chemical reactions. The purpose of this work is to advance the concept of the Directly Irradiated Fluidized Bed Autothermal Reactor (DIFBAR), a newly-designed solar reactor, that exploits fluidization technology and the principle of an autothermal reactor: the thermal energy of the solids flowing out of a solar receiver is used to preheat the feed in a countercurrent double-pipe heat exchanger. This allows to reach high temperatures with minimum solar power input, reducing the size of the CST facility. A full DIFBAR prototype has been tested with a high-flux solar simulator, consisting of an elliptical reflector and a 10 kW_{el} Xe arc lamp. The prototype is operated as a circulating fluidized bed: the inner tube of the heat exchanger is a fluidized bed riser, the receiver works as a gas-solid separator, the outer tube (annulus) of the heat exchanger is an overflow standpipe and a buffer tank (reservoir) connects the annulus to the riser, closing the loop. The reservoir can also be operated as a fluidized bed reactor like in a dual-fluidized bed system. Experiments confirm the viability of the autothermal operation. Results, obtained for different bed materials (Geldart B class), cover all the fundamental aspects of the system: hydrodynamic control, thermal behavior and chemical reaction. Solids circulation rates up to 9 kg/h are measured at ambient temperature, while varying the gas velocity in the riser. Heating tests are carried out for different irradiation powers. Steady state temperature profiles are analyzed to assess the performance of the heat exchanger: the heat transfer coefficient is evaluated between 340 and 490 $W/(m^2\ K)$. Preliminary tests involving the calcination of magnesium and calcium carbonate are carried out to investigate the operability of Chemical Looping processes.

Effect on Material Properties after Cyclic Fluidized Bed Reduction of Iron Ore Fines

Authors

Stevens, N. - Presenter

Deen, N., Eindhoven University of Technology

Finotello, G., Eindhoven University of Technology

Recently iron powder has been proposed as a high energy density, easily storable, and CO2-free energy carrier. During the combustion of iron powder with air energy is released and then the combusted products, the iron oxide particles, are captured and cyclically reduced back into iron powder via an advanced technology that is powered by renewable energy. A combustion step with a subsequent reduction corresponds to one cycle. In this work we focus on experiments of hydrogen reduction of iron oxides under fluidization conditions and we determine the potential of the entire iron cycle, in terms of the material properties. The powder is combusted and reduced for three cycles under the same conditions and characterized after each cycle process. The physical techniques used for size and shape analysis includes laser diffraction particle size analysis and scanning electron microscopy. For structure and chemistry X-ray diffraction and energy dispersive X-ray analysis are also employed.

To study the effect of the reduction conditions on the process and powder, two cyclability tests of three cycles are conducted at 550 °C and 575 °C as reduction temperatures. These temperatures are selected to identify the trade-off between a high reduction degree and absence of particle sticking. We observed that the particle size distribution is approximately the same over the three cycles, both at 550 °C and 575 °C, indicating that the powder can be used in the cyclic process without intermediate treatments (e.g. sieving). However, at 575 °C sticking negatively affects the fluidization and the reduction reaction (average RD is 4% lower than at 550 °C) We observed that the composition of the reduced powder after each cycle consists mainly of iron and partly magnetite. When looking at the porosity, the results show that the pores size of the particles increases with temperature.

Modeling and Optimizing Gas Solid Distribution in Fluidized Beds

Authors

Singh, R. - Presenter, Technip Energies Process Technology

Shimoda, S., Technip Energies Process Technology

Golczynski, S., Technip Energies

In fluidized bed reactors, uniform distribution of gas and solids is critical to achieve the desired hydrodynamic and kinetic performance. For FCC units, Technip Energies uses a proprietary distributor design to terminate both reactor risers and spent catalyst lift lines to distribute the exiting gas-solid stream into a fluidized bed.

In Technip Energiesâ€™ Resid FCC unit, air and spent catalyst is distributed from an upward flowing stream into catalyst bed for coke combustion on catalyst. Proper distribution is key to achieve a uniform coke burn and an even temperature profile throughout the bed. In PropyleneMax™ Catalytic Cracking (PMcc™) technology, the riser reactor also terminates in a fluidized bed, requiring proper distribution of the hydrocarbon vapor and solid mixture into the reactor bed, which is key to providing adequate contact for further cracking to valuable products. This paper discusses Technip Energiesâ€™ gas-solid distribution technology and its recent work to develop an improved distributor for use as a riser termination device in both the PMcc reactor and the RFCC regenerator to achieve improved performance.

Use of CFD modeling to screen and evaluate the initial concepts and optimize the final selected concept will be presented. Furthermore, CFD results from both reactor and regenerator side will be discussed with a new distributor design showing benefits of improved gas solid distribution on hydrodynamic and kinetic performance. The performance matrix used to quantify the benefits of the optimized design will be highlighted

For regenerator modeling, coke burning kinetic description is provided through proprietary combustion kinetic equations. The results provide detailed qualitative mapping of the regenerator behavior in terms of flue gas composition, temperature distribution including afterburning, regenerator bed density in both axial and radial planes. The CFD coupled with spent catalyst regeneration kinetics is used to quantitative evaluate regenerator performance and benefits of technology upgrade

Subgrid Drag Model Derivation from Drift Velocity Equation for Coarse-Grid Euler-Euler Simulations of Fluidized Beds

Authors

Simonin, O. - Presenter, Université de Toulouse, CNRS-Toulouse

Hardy, B., Université Catholique de Louvain

Fede, P., Institut de Mécanique des Fluides de Toulouse

The industrial-scale simulations of fluidized bed reactors are currently performed using a Two-Fluid modeling approach where both the gas and the particles are treated as inter-penetrating continua. Yet, the size of the mesh used is usually larger than the size of the mesoscale structures that appear in such gas-solids flows, namely clusters of particles. Omitting the influence of these subgrid scale structures on the resolved scales of the flow was shown to give poor predictions of the bed hydrodynamics. The evaluation of the subgrid scale fluid-particle drag term currently relies on the modeling of the so-called drift velocity, due to the correlation between the solid volume fraction and the fluid velocity at the subgrid scale. Functional models based on the resolved particle volume fraction and the size of the coarse grid have been proposed to estimate the drift velocity from resolved quantities. In this study, a novel modeling approach is proposed. First, a transport equation is derived for the drift velocity. Fine-grid simulations of a tri-periodic fluidized bed reactor are performed to obtain mesh-independent results and the data are spatially filtered to perform *a priori* analyses of the coarse-grid equations. It appears from a budget analysis that the term involving the fluid-particle drag force is dominant in the drift velocity transport equation. An explicit model is derived where the drift velocity is expressed as a function of the resolved slip velocity and an algebraic function of a few moments of the solid volume fraction. Fine-grid simulations also show that the solid volume fraction conditioned by its filtered value and its subgrid variance follows a beta-type distribution. Henceforth, the proposed drift velocity model relies only on the local filtered volume fraction and its variance, for which a distinct transport equation or an algebraic model can be used.

Modeling of Fluid Turbulence Modulation and Pseudo-Turbulence Production By Particles in Liquid-Solid Fluidized Bed

Authors

Ansart, R. - Presenter, Université de Toulouse, CNRS-Toulouse

Simonin, O., Institut de Mécanique des Fluides de Toulouse

Ritta, A. S. L., Université de Toulouse

Numerical modeling of fluidized bed processes from lab scale to industrial scale by N-Euler CFD modeling approaches is becoming an essential tool to support the optimization or scaling-up of existing processes and the development of new processes based on innovative concepts. The liquid-solid fluidized beds are used in numerous processes, including chemical, biochemical and catalytic reactions. To gain an understanding on this complex industrial process requires a greater knowledge of the multi-scale hydrodynamic interactions that exist between the particle and the fluid phases.

A new two-way coupling model is proposed to account for both the effect of particles on the large scales of the fluid turbulent motion and production of velocity fluctuations in the wakes of the particles. This approach is based on separate equations and particle modulation terms for large-scale turbulence Reynolds stresses and for pseudo-turbulence kinetic energy. This modelling approach will be implemented in neptune_cfd code and apply to simulate a 3D liquid-solid fluidized bed. The simulation results are going to be compared with experimental results from Almeras et al. (2021). This new mathematical modeling will increase the fluid velocity fluctuation prediction due to the contribution of the pseudo-turbulence without relevant impact of prediction on particle velocity fluctuation.

In addition, a study will also be presented on the effects of different types of rheological models of the liquid phase: either homogeneous as a function of the solid volume fraction or as a function of the local pseudo-turbulence. The impact of these models on the spatial and time velocity variances of the fluid will be investigated.

Application of Filtered Two-Fluid Models to Industrial-Scale Fluidized Beds

Authors

Vandewalle, L. A. - Presenter, Ghent University, Laboratory For Chemical Technolo

Buettner, K., Exxonmobil

Healy, T. M., ExxonMobil Research and Engineering Company

Supported by recent developments in high performance computing capabilities, computational fluid dynamics (CFD) has become an invaluable tool in the study, design and scale-up of gas-solid fluidized beds. In an industrial context, CFD of commercial-scale fluidized beds is used either in the development stages of new processes and equipment, or as a way of troubleshooting and improving existing production assets.

Two approaches are typically considered for modeling gas-solid flows. While the Eulerian-Lagrangian approach (e.g., CFD-DEM) is increasingly used for fundamental studies in academic research, it is still too computationally expensive for industrially relevant scales in refining and chemicals applications. The preferred method for industrial applications is therefore the Eulerian-Eulerian approach (i.e., two-fluid model). The two-fluid model with classical drag laws is able to capture the complex multi-scale flow phenomena occurring in gas-solid flows, but to preserve accuracy, the mesh resolution should be limited to ten particle diameters at most [1], which is impractical for most full-scale applications. By replacing the classical micro-scale drag laws with a meso-scale â€˜filteredâ€™ model, the mesh resolution requirements and hence computational cost can be significantly reduced [2]. These filtered models are derived from highly resolved two-fluid simulations in a periodic domain, which are then filtered using various filter sizes, with the ultimate objective of producing an accurate drag force in cells that are larger than 100 particle diameters.

The CFD group at ExxonMobil has been using filtered drag models for over a decade. In this presentation, industrial successes in applying these models to refining and chemicals processes will be discussed. Furthermore, we will highlight some challenges that had to be tackled when using the existing filtered models outside their original validation range.

[1] Agrawal, K. et al. Journal of Fluid Mechanics 445 (2001) 151-185.

[2] Igci, Y., et al. AIChE Journal 54 (2008) 1431-1448.

Experimental Analysis of a Confined Bed of Granular Material As Thermal Energy Storage System

Authors

Soria-Verdugo, A. - Presenter, University Carlos III of Madrid

Hernández-Jiménez, F., University Carlos III of Madrid

García-Gutiérrez, L. M., University Carlos III of Madrid

Cano-Pleite, E., University Carlos III of Madrid

García-Hernando, N., University Carlos III of Madrid

Felix Guil-Pedrosa, J., University Carlos III of Madrid

Thermal Energy Storage (TES) systems contribute to the uniform and efficient production of green energy in concentrated solar power plants. Storage of sensible heat in a granular material is widely used, among the different alternatives of TES systems, due to its simple operation. In this work, a lab-scale TES system is experimentally analyzed. The setup consists of a cylindrical vessel of stainless steel with 20 cm in diameter and 1 m in height, filled with sand as granular material, using air as working fluid. A perforated plate distributor prevents the sand particles to fall in the plenum chamber at the bottom of the vessel. The vessel walls are surrounded by 4 electric heaters to heat up the system to the desired initial temperature. The system is equipped with a pressure sensor at the plenum chamber to measure the pressure drop of the bed, and with thermocouples in the plenum chamber and at the outlet of the bed to determine the inlet and outlet temperature of air during the discharge process. Thermocouples are also distributed along the bed height in intervals of 5 cm to measure the evolution, with time and bed height, of the temperature at the center of the bed. The discharging process of the system is experimentally characterized using air velocities below and above the minimum fluidization velocity of sand, operating the system in fixed and fluidized bed regimes. For the case of velocity higher than the minimum fluidization velocity, another perforated plate distributor was used at the top of the vessel to mechanically confine the bed, preventing the motion and thus the fluidization of the particles. The distribution of the bed temperature for the fixed, fluidized, and confined beds are compared in terms of the air exergy availability at the outlet of the TES system.

Employing Positron Emission Tomography to Study the Dynamics of Droplets in Fluid Flows

Authors

Offner, A. - Presenter

Positron emission tomography (PET) is a method used to diagnose and map the spread of cancer in human body using radioactive agents. In this method, the radioactive material spontaneously emits positrons that undergo annihilation, thus generating a pair of photons that are fired back-to-back to conserve momentum. The high-energy photons penetrate nearly any medium (including most metals) and are sensed on a grid of small detectors (scanner), allowing for triangulation of the annihilation position and ultimately for the reconstruction of the three-dimensional radioactivity distribution in the entire body. In the present work, we leverage on the strengths of PET and exploit it to study the dynamics of droplets suspended in fluid. We produce droplets from Fluorine-18 (^{18}F) water and release them in a vessel filled with an Ethanol-Glycerol binary mixture, positioned inside a PET scanner. By changing the mixture ratio, we tune its density and the diffusivity of water in it, allowing us full control over the experiment pace which in turn ensures the dynamics is captured with the radioactivity levels we used. Using a unique probabilistic model we developed, we infer the droplet trajectories and their time-dependent activities, that is their rates of positron emissions, which can accurately estimate their size evolution. Our inference is compared against results from image analysis to video footage of the experiment, showing good quantitative agreement. The use of PET to study the fluid dynamics of suspended droplets can potentially be used in a variety of applications, including drug delivery or emulsion stability in flows.

Hydrodynamic Characterization of Different Geldart B Solids in Dense Gas-Fluidized Beds By Capacitance Probes

Authors

Molignano, L. - Presenter

Troiano, M., Università degli Studi di Napoli Federico II

Solimene, R., Consiglio Nazionale delle Ricerche

Tebianian, S., IFP Energies nouvelles

Joly, J. F., IFPEN

Salatino, P., Università degli Studi di Napoli Federico II

Fluidization technology presents valuable options for efficient thermochemical processing of solid fuels. In this framework, bed hydrodynamics strongly influence the performance of the reaction as they determine mixing/segregation patterns of the fluidizing gas, and of the fuel in the inert fluidizing medium. Knowledge about the last-mentioned aspects is a key prerequisite for successful design and operation of such converters. However, there are still broad areas of uncertainties regarding fluidized beds hydrodynamics, mixing, heat and mass transfer phenomena.

In the present study, the hydrodynamics of bubbling fluidized beds operated at ambient temperature and 500°C are investigated by capacitance probes at different gas superficial velocities. The probes are fitted to a medium-sized laboratory scale unit with the sensing volume placed at multiple radial/axial column coordinates. Statistical analysis of the time-series of the local bed voidage generates probability density functions that display multimodal patterns. A bimodal character can be ascribed to the emulsion phase when the gas superficial velocity is increased. It is inferred that two phases with different porosity co-exist in the emulsion phase: a lower voidage LV-phase and a higher voidage HV-phase.

Three granular materials belonging to group B of Geldart classification are investigated, which differ for particle diameter, density and sphericity. Quantitative and qualitative differences in terms of hydrodynamic behaviour are highlighted and related to the above-mentioned solids properties.

The influence of the bed hydrodynamics on the gas mass transfer around freely moving coarse active particles is also assessed. Based on the results obtained by the above-described experimental campaign, the use of a Frössling-type equation is critically reviewed in the light of the observed expansion patterns of the emulsion phase.

Mitigation of Electrostatic Charge and Its Adverse Effects in Gas-Solid Fluidized Beds

Authors

Mehrani, P. - Presenter, University of Ottawa

Sowinski, A., University of Ottawa

Isaac Nimvari, M.

Sridhar, N., University of Ottawa

In gas-solid fluidized beds there is potential for electrostatic charge generation due to the collisions among particles and with the column wall. Operational challenges arising from electrostatics include particle agglomeration, accumulation of charged particles on internal surfaces (column wall, instrumentation, etc.,) requiring shutdowns for clean-ups, and electrostatic discharge resulting in a hazardous environment. Previously we have shown that the extent of charging largely depends on various conditions including the nature of surfaces in contact (i.e., surface chemistries of particles and the column wall), gas velocity and flow regime, presence of moisture, fluidization pressure and temperature, and particles size distribution [1,2]. Our latest works indicate that there is a direct relationship between the net specific charge of the particles in the bed bulk and the degree of particlesâ€™ accumulation on the fluidization column wall [2,3]. This presentation presents three techniques assisting in minimizing the generation of electrostatics in gas-solid fluidized beds. They include: (1) Reducing volume resistivity of the particles (e.g., via increase in fluidization temperature); (2) Reducing the particles surface charge/charge build-up using fluidization gasses that have low breakdown voltage strength; and (3) Adding additives (e.g., antistatic agents) that induce an opposite polarity to that of the net specific charge of the bed bulk, and in turn, bringing the bed net specific charge as close as possible to zero.

[1] Mehrani P., Murtomaa M., Lacks D.J., An overview of advances in understanding electrostatic charge buildup in gas-solid fluidized beds, Journal of Electrostatics, 87, 2017.

[2] Nimvari, M.I., Sowinski, A., Mehrani, P., Effect of Temperature on triboelectrification of polyethylene particles in a pilot-scale pressurized gas-solid fluidized bed. Powder Technology, 405, 2022.

[3] Taghavivand, M., Mehrani, P., Sowinski, A. Triboelectric effects of continuity additives and a silica catalyst support on polyethylene fluidized bed wall fouling, Chemical Engineering Science, 245, 2021.

Experimental Study on Vibro-Fluidized Bed Drying

Authors

De Munck, M. - Presenter, Eindhoven University of Technology

Marrevee, D., Eindhoven University of Technology

Peters, E. A. J. F.

Kuipers, J. A. M., Eindhoven University of Technology

Vibro-fluidized bed drying is commonly encountered in several industries in order to reduce the moisture content in products. Drying is a continuously evolving process where the particle and gas properties change over time. The particle mass reduction and the gas moisture saturation are key factors that influence the process properties such as the fluidization behavior. Conversely, the changing fluidization behavior influences the mass and heat transfer characteristics. Vibration enhances the fluidization characteristics, hence solids drying is improved compared to static fluidized beds.

In this study, experiments in a pseudo-2D vibro-fluidized bed setup are performed in order to better understand the improved solids drying behavior in vibro-fluidized beds. A coupled Particle Image Velocimetry - Infrared Thermography technique is employed to characterize the local solids velocity and temperature fields. In the first part, the effect of mechanical vibration on monodisperse solids drying is discussed. The second part focuses on the extension to binary and ternary mixtures where a machine learning algorithm is applied to characterize the size-based mixing and segregation dynamics. It is found that size-based segregation is suppressed under the influence of mechanical vibration.

Numerical Study of the Erosion of an Immersed Tube in a Bubbling Fluidized Bed

Authors

Domínguez, P. - Presenter

Corcoles Tendero, J. I. - Presenter

Almendros Ibáñez, J. A., Universidad de Castilla-La Mancha

Díaz-Heras, M., Castilla - La Mancha University

The erosion of internals in fluidized beds is a problem that still holds the attention of researchers. The main reasons are the complexity of the process itself and the difficulties in defining the variables involved in the process. The multiple possible situations depending on the geometry, particle characteristics and fluidization parameters embiggens the problem.

Several experimental and numerical studies have analysed the erosion on immersed tubes in fluidized beds. Most of the published papers during the 70â€™s, 80â€™s and 90â€™s were focused in fluidized bed combustors and there is no studies on new fluidized bed applications, such as Concentrating Solar Power (CSP) applications, with novel particles as SiC, which are very hard and sharp. Regarding numerical simulations, it is highlighted the use of Multiphase Particle in Cell Method (MP-PIC), which considers group of particles with the same properties instead of individual particles.

The aim of this paper is to perform a numerical simulation of a 2-D single tube in a fluidized bed in Barracuda Virtual Reactor (v. 22.1) to predict the erosion in the perimeter of the tube. The numerical results are compared with previous studies (Xu et al. in 2022). Different grid size and time steps are analysed.

The numerical results clearly shows that the erosion of the tube is not homogeneous along the circumference of round tubes, being higher on the lower lateral areas of the tube and minimum on the top, where the particles are at rest. The plot of the erosion versus the circumferential position typically shows an inverted W-shape, with a relative minimum at the tube bottom (although the erosion remains high).

Surface Morphology of Granules Coated with Different Materials from Fluidized Bed Spray Granulation

Authors

Orth, M. - Presenter

Haschimi, M., Hamburg University of Technology

Pietsch-Braune, S., Hamburg University of Technology

Heinrich, S., Hamburg University of Technology

Fluidized bed spray granulation is a versatile process that can be used for production of a broad variety of particulate products, for example in the pharmaceutical, food, and chemical industry. Along with the possibility to produce such tailor-made granular materials comes the challenge to adjust the process in such a way that required product properties are met when working with different materials. During granulation, the structure formed from primary particles and injected coating material is determined by the interaction of the three phases involved in the process and various micro mechanisms occurring due to these interactions. Depending on the materials used in the process, different phenomena are decisive for the formation and final structure of the coating layer.

In this work, granulation experiments were carried out in a lab-scale fluidized bed with Cellets®500 as core particles. As coating material, sodium benzoate and calcium carbonate were used and injected in form of an aqueous solution and suspension, respectively. For both materials, experiments were performed at varied process parameters to investigate how these changed conditions influence the coating layer formation. In order to analyze the surface morphologies, the surface roughness of the granules was measured using confocal laser-scanning microscopy. Additionally, the coating layer around the Cellet®500 core was characterized regarding its thickness using SEM images of cut particles as well as particle size distribution measurements. In case of sodium benzoate as coating material, layer structure seemed to be predominantly dependent on the crystallization of the salt on the particle surface. In contrast, in the suspension-based coating other micro mechanisms like droplet spreading have a more pronounced impact on surface morphology.

Charge-Velocity Correlation Transport Equation Model with Triboelectric Effect for Gas-Solid Fluidized Bed N-Euler Simulation

Authors

Simonin, O. - Presenter, Institut de Mécanique des Fluides de Toulouse

Montilla, C., Laboratoire de Génie Chimique, Université de Toulouse

Ansart, R., Université de Toulouse, CNRS-Toulouse

The modelling and simulation of gas-solid fluidized bed is complex due to the multi-physical nature of particle-laden flows. A non-exhaustive list includes particle-turbulence interaction, inter-particle collisions or particle bouncing on smooth or rough walls. In addition, although long neglected in modelling approaches, the effect of electrical charges on the macroscopic behavior of gas-solid fluidized can be very important. The objective of this paper is to present some recent advances in modelling and simulation of electrically charged particles with triboelectric effect in gas-solid fluidized beds.

The kinetic theory of granular flows is used to derive the separate transport equations for the mean particle electric charge, the chargeâ€"velocity covariance and the charge variance for monodisperse particles in gas flow with tribocharging effect. The collision terms are closed by assuming that the electrostatic interaction does not modify the collision dynamics and without presuming explicitly the form of the dependence of the joint charge-velocity PDF on the particle electric charge. The full second-order transport equation model is tested in a one-dimensional periodic domain with nonuniform initial mean charge distribution. The results show that this model is able to capture important physical mechanisms that are neglected by simple algebraic models proposed for the particle charge-velocity correlation derived in previous works. An in-depth analysis of the transport equations is also performed. This study reveals that, for sufficiently small covariance characteristic destruction time scales, the transient and third-order moments terms can be safely neglected. In addition, reduced-order models are derived and analyzed. In particular, a semi-algebraic model that only resolves a transport equation for the charge variance coupled with an algebraic model for the covariance is proposed. Such a model is found to be a suitable alternative for sufficiently small value of a dimensionless interparticle collision time.

The Fragmentation-Induced Fluidization of Pyroclastic Density Currents

Authors

Breard, E. - Presenter, Georgia Institute of Technology

Dufek, J., University of Oregon

Charbonnier, S., University of South Florida

Gueugneau, V., University of South Florida

Giachetti, T., University of Oregon

Pyroclastic density currents (PDCs) are the most lethal volcanic process on Earth. Forecasting their inundation area is essential to mitigate their risk, but existing models are limited by our poor understanding of their dynamics. Here, we explore the role of evolving grain-size distribution in controlling the runout of the most common and deadliest PDCs known as block-and-ash flows (BAFs). Through a combination of theory, analysis of deposits and experiments of natural mixtures, we show that rapid changes of the grain-size distribution transported in BAFs result in the reduction of pore volume (compaction) within the first kilometres of their runout. We then use a multiphase flow model to show how the compressibility of granular mixtures leads to fragmentation-induced fluidisation (FIF) and excess pore-fluid pressure in BAFs. This process dominates the first ~2 km of their runout, where the effective friction coefficient is progressively reduced. Beyond that distance, transport is modulated by diffusion of the excess pore pressure. Fragmentation induced fluidisation provides a physical basis to explain the decades-long use of low effective friction coefficients used in depth-averaged simulations required to match observed flow inundation.

Influence of Sparger Design on Local Phase Volume Fraction Distribution in a Slurry Bubble Column: ERT Measurements and Eulerian CFD Simulations

Authors

Manna, S. - Presenter

Saini, R.

Buwa, V., Indian Institute of Technology Delhi

Gas-liquid-solid bubble column reactors (SBCRs) are used to perform solid-catalysed gas-liquid reactions like FT synthesis, resid-hydrocracking etc. Sparger design significantly influences gas/solid phase distribution and thus the performance of a slurry reactor. In past, though the effect of sparger design on â€œoverallâ€ gas hold-up was investigated, its effect on local â€œspatio-temporalâ€ distribution of gas and solid volume fraction is not investigated. Thus, the main objectives of the present work are (1) to understand the influence of different sparger configurations through space- and time-resolved measurements of gas volume fraction distribution and (2) to perform corresponding Eulerian CFD simulations to understand the influence of sparger design on space- and time-resolved gas/solid phase distribution and the dynamics of dense three-phase flow in a bubble column.

The effect of sparger designs (uniform, 4/2/1 holes) is investigated by measuring local gas volume fraction distribution in a cylindrical column (d_C=0.19m, h_c=1.35m) using Electrical Resistance Tomography (ERT) for water-air-alumina system for solid loading of 0, 10 vol% and superficial gas velocity (U_G) in the range of 5-30 cm/s, under ambient conditions. Due to significant settling of solids, gas was found to rise in the near-wall region in the sparger-zone at U_G=5cm/s, while uniform gas distribution was observed at U_G=30cm/s. A CFD model is developed based on three-fluid Eulerian+ KTGF approach. We have investigated the effect of drag force with bubble swarm corrections, and currently we are investigating the effects of non-drag forces like lift and turbulent dispersion on the local gas distribution. This model will be validated against measurements and will be used for predictions of phase distribution and dynamics of three-phase flow in SBCRs for different sparger designs. The space-/time-resolved measurements performed using ERT and development of experimentally-validated CFD models will be an important step forward in predicting the performance of slurry reactors for commercial applications.

CFD-DEM Modeling of Laboratory-Scale Fluidized BED Containing Binary Mixtures: Validation with Radioactive Particle Tracking (RPT) Experiments

Authors

Nath, S. - Presenter

Roy, S., Indian Institute of Technology Delhi

Gas-fluidized beds are extensively utilized in industrial processes owing to the fact that these systems offer advantages like uniform particle mixing, possibility to operate in continuous state, uniform temperature gradients. Due to complex hydrodynamics and lack of complete understanding of these systems, accurate scale-up of such units is challenging. Therefore, gaining more insights into the complex hydrodynamics is of utmost importance and this can be achieved through development and validation of underlying hydrodynamic models. Eulerâ€"Lagrange model/ Discrete Particle Model arguably gives more physically realistic results, as this method is based on the tracing of each particle inside the system. The present study seeks to capture these complicated flow patterns at various conditions using Euler-Lagrange approach.

CFD-DEM simulation of a laboratory-scale 3D binary fluidized bed has been carried out consisting of glass beads of 0.5- and 2-mm diameter particle sizes. To determine the contact forces between particles, Hertzian model was used. Effect of bed composition was checked on overall hydrodynamics of the system by varying the coarser particle fraction from 10 â€" 40% at different superficial gas velocities. Effect of different drag models like Gidaspow, Di Felice and Koch-Hill were also observed in the simulations. For validation of the efficacy of the simulations, a direct comparison is made of the results obtained, i.e., minimum fluidization velocity, time averaged axial mean velocity and RMS velocity of solid, with an experimental technique that is Lagrangian in nature, *viz*. Radioactive Particle Tracking (Roy et al., 2021). The observations from simulation include not only comparisons of the mean profiles, but also the fluctuating components of velocity from the flotsam and jetsam phases. As the bed consists of a distribution of particle sizes, the mixing pattern of the polydisperse bed is also studied.

Keywords: Computational Fluid Dynamics, Discrete Element Model, Binary Fluidized Bed, Radioactive Particle Tracking

The Development of Improved Particle Drag Force Correlations for CFD Simulation

Authors

Parker, J. - Presenter, CPFD Software

The particle drag force correlation is often the most important factor in a fluidized bed CFD simulation. While commonly-used drag models capture the basic functional dependencies on Reynolds number and voidage, the real-world particle dynamics may also be affected by particle characteristics such as the particle shape, the full-size distribution, and surface properties. These additional factors, which are often difficult or expensive to measure, can be a source of error in the particle drag force correlation and a CFD simulation as a whole. Therefore, in common practice, the accuracy of a particle drag force model is improved through iterative tuning based on experimental bed density, entrainment rates, or other fluidization characteristics.

The goal of this work is to reduce the computational time and cost of CFD tuning by developing particle drag force correlations which can be directly improved based on easily-obtained experimental pressure drop or minimum fluidization data. Model accuracy before and after tuning is demonstrated using the wide range of liquid-solid bed expansion and settling data that is available in literature. The same data sets are used for comparison with particle drag force models that commonly used in CFD.

Physical Modelling and Hydrodynamical Characterization of an Internally Circulating Fluidized Bed for Biomass Fast Pyrolysis

Authors

Troiano, M. - Presenter, Università degli Studi di Napoli Federico II

Padula, S., Federico II University

Solimene, R., Consiglio Nazionale delle Ricerche

Salatino, P., Università degli Studi di Napoli Federico II

Biomass represents a promising renewable source for the substitution of fossil feedstocks in the production of fuels and platform chemicals. Biomass fast pyrolysis offers a direct route to produce liquid fuels and chemicals, with high feedstock flexibility and energy input efficiencies. Fast pyrolysis is characterized by operation at a moderate temperature, short residence time of the vapor and high heating rate. The liquid products offer significant advantages in storage and transport and for this reason they can be used as biofeedstock in a biorefinery scenario.

Critical issues of biomass fast pyrolysis are the heating rate, the raw biomass feeding point, the mixing of the solid fuel with the inert bed material, the contact time between char and vapours. Furthermore, residence times have to be carefully limited below 1s to avoid secondary homogeneous and heterogeneous pyrolysis reactions and the fluid dynamic behaviour of both solid and gaseous phases should be accurately controlled on the basis of the desired reactive pathways.

This study addresses a hydrodynamical characterization of an internally circulating fluidized bed reactor for the fast pyrolysis of biomass. Inert particles are continuously circulated between a fluidized, and a moving bed section, divided by a vertical baffle and interconnected by an opening at the bottom. Particles in the fluidized bed section are transported upwards over the baffle to the moving bed section, where they flow down by gravity and return in the fluidized bed across the opening. The hydrodynamic behaviour of the internally circulating fluidized bed is experimentally investigated in a cold flow model apparatus. Experiments aim at studying the driving force and the mechanism of particle circulation in the system. Experimental tests allow the measurement of solids circulation rate and the assessment of particle residence time distributions under different operating conditions and for different opening ratios.

Surface Waves and Pressure Fluctuations in Fluidized Beds with an Oscillating Plate

Authors

Bi, X. - Presenter, University of British Columbia

Li, J., China University of Petroleum (Beijing)

Niu, L., China University of Petroleum, Beijing

Jia, D., University of British Columbia

Surface waves generated by a horizontal oscillating paddle at the upper surface of non-shallow fluidized beds were studied by Finnerty et al. (1969). The wavelength and attenuation-rate were measured from motion pictures. Measured wavelength and attenuation followed the trends of the dynamic model for pseudo-liquid with an effective viscosity, with the attenuation rate increased with increasing wave frequency. Similar experiments were performed by Kok and Benschop (1994) by measured pressure fluctuations. The wave propagation velocity and attenuation rate with frequency <1.6 Hz were found to follow the trends as predicted by the pseudo-liquid model. Beyond 1.6 Hz, the trends became opposite to those predicted from the surface wave theory, attributed to a transition from surface waves to density waves due to the off-design operation of the oscillating paddle.

Both studies used oscillating paddle surface wave generators. While Finnerty et al. measured the surface wave attenuation rate directly from recorded motion pictures of surface waves, Kok and Benschop determined the wave properties indirectly from measured local pressure fluctuations. because local pressure fluctuations in gas-solids fluidized beds are composed of multiple sources (see Bi, 2007) including global bed oscillations, local bubble passage, and propagating pressure waves originating in other locations, there is a need of further experiments to understand the difference in the two measurement techniques and the transition around 1.6 Hz in Kok and Benschopâ€™s experiments.

A cold model fluidized bed was constructed with a variable-speed motor-driven oscillating plate and multiple pressure transducers at different locations of the bed. FCC and glass beads particles of Group A and Group B, respectively, were used as the bed material. Wave frequency, propagation velocity and attenuation rate were measured from pressure fluctuations at different locations. Key findings on the complex local pressure fluctuations and their relationship with the surface waves are to be presented.

Simulation of an Autothermal Steam Oxygen-Bubbling Fluidized Bed Gasification of Biomass Waste for Hydrogen and Biofuel Production

Authors

Anaya, B. - Presenter

Materazzi, M., University College London

Developing a predictive model of a bubbling fluidized bed reactor is challenging due to the complexity of gas-solid mixing patterns and the existence of a broad range of volume- and temperature-sensitive reactions. This study presents a predictive model of an autothermal bubbling fluidized bed gasifier operated with steam and oxygen that utilizes waste biomass to produce clean syngas, suitable for synthesis of biofuels and hydrogen by using process simulation software, such as Aspen Plus. The two-phase theory of fluidization's underlying principles were taken into account when designing the model. The reactor is divided into three main reacting zones connected in series and parallel: the dense/bed phase (emulsion plus bubbles), the splashing zone where the feed entry is located, and the freeboard. Using experimental, pilot-scale data, the model was validated with high accuracy. This model can be employed to assist industries with process design and scale up, and further improve the process efficiency and/or product quality for bubbling fluidized bed gasification of various biomass feedstock types

AUTHOR INDEX

AIChE
120 Wall Street, FL 23
New York, NY 10005-4020

ISBN 978-1-7138-8677-8

9 781713 886778

CURRAN ASSOCIATES INC.
proceedings
.com